SpringerBriefs in Biochemistry and Molecular Biology

AF212061

More information about this series at http://www.springer.com/series/10196

KBBE Synthetic Biology

ST-FLOW

FP7 project

Manuel Porcar · Juli Peretó

Synthetic Biology

From iGEM to the Artificial Cell

 Springer

Manuel Porcar
General Foundation, Cavanilles Institute
  for Biodiversity and Evolutionary
  Biology
University of Valencia
Paterna
Spain

Juli Peretó
Department of Biochemistry and Molecular
  Biology, Cavanilles Institute for
  Biodiversity and Evolutionary Biology
University of Valencia
Paterna
Spain

ISSN 2211-9353
ISBN 978-94-017-9381-0
DOI 10.1007/978-94-017-9382-7

ISSN 2211-9361 (electronic)
ISBN 978-94-017-9382-7 (eBook)

Library of Congress Control Number: 2014947696

Springer Dordrecht Heidelberg New York London

Printed on acid-free paper

Springer is part of Springer Science+Business Media (www.springer.com)

To the Synthetic Biology pioneers, particularly the young igemites we had the chance to work with.

Foreword

Extant biology continuously undergoes rapid transformation, with a proliferation of new disciplines: systems biology, epigenetics, proteomics, synthetic biology... Among these new developments, synthetic biology is certainly the most significant. That is why this book, by Manuel Porcar and Juli Peretó, is most welcome. Despite its modest length, it is rich in information and reflections on this new domain of research.

The authors outline the main developments in synthetic biology since it appeared, at the beginning of the twenty-first century, up until the present day. This is by far the best way to provide a comprehensive picture of this new discipline. The authors distinguish different experimental approaches, bottom up, top-down, and xenobiology. They describe the most visible works (such as J. Craig Venter's projects) as well as less publicized—but equally important—contributions. Drawing on their own experience, they describe the famous iGEM student competition. Not only does this annual event promote the visibility of synthetic biology and make it more appealing, but iGEM also illustrates its chief characteristics.

Peretó and Porcar have also devoted a large part of this book to history, illustrating how synthetic biology is the legacy of a long tradition of research. They describe the early twentieth century experiments of Alfonso L. Herrera in Mexico, and Stéphane Leduc in France; scientists who were already endeavoring to synthesize life. Indeed, it was Leduc who coined the expression "Synthetic Biology" back in 1912. During the same period, more important than these illusory accomplishments, was the construction of a conceptual framework by authors like Jacques Loeb, which facilitated the subsequent development of synthetic biology.

This historical sketch also shows that the media hype surrounding the "creation of life" is not new, but has appeared time and again during the past century.

This book is also a source of reflection on the significance, and the bounds, of the new discipline. One interesting issue is the comparison between organisms and machines, initially made by Descartes. It is remarkable how dissimilar the present attitudes of American and European biologists are. The former do not reject the comparison, whereas the latter take a much more cautious view. The promiscuity of the parts within organisms—their capacity to accomplish different functions—is a property that machine components do not share. One reason for this divide between organisms and machines is that the former are the result of a long evolutionary history.

Maybe the most important message conveyed in this book, and the main reason why synthetic biology should be considered so important, relates to what it can teach us about life. Successful experiments modifying extant organisms, or synthesizing new artificial life forms, will be the best proof of the value of the knowledge we have gathered about life over the decades. Even though some projects may be considered failures, they will help biologists to modify their conception of organisms, and the "definition of life."

Besides the practical prospects of synthesizing biofuels and new drugs or replacing dirty chemistry with clean processes, synthetic biology is the best path toward exploring the "mystery of life." This book is a wonderful introduction to this enterprise!

Michel Morange
Professor of Biology at École Normale Supérieure (ENS, Paris),
Université Pierre et Marie Curie (Paris 6),
Director of Centre Cavailles for History
and Philosophy of Sciences (ENS, Paris)

Preface

In his enlightening *Letters to a Young Scientist* (2013) myrmecologist and socio-biologist Edward O. Wilson beautifully describes three different archetypes of the scientific mind. The creative life of a scientist could be oriented to a travel into unexplored regions (e.g., the detailed molecular cartography of a cell or the most remote galaxies), a fight against evil (e.g., the great health afflictions or the shortage in energy sources), and a kind of Grail Quest. Among the extraordinary search of a real grail Wilson proposes "the creation in the laboratory of a simple organism." It is hard not to agree that this experiment could be the most intellectually shocking in human history: a second example of life in our hands.

As scientists involved in research and teaching in diverse areas of biology, such as biochemistry, genetics, or biotechnology, we are also engaged in the development of synthetic biology, a multifaceted approach to redesign living organisms. In particular, we have been involved in the development of several student projects for the international competition in synthetic biology iGEM, and have joined the debate on the nature life, its origin, and its possible synthesis in the laboratory. Actually, the origin of our adventure of writing this book started with a paper of ours entitled "Are we doing synthetic biology?"

We are witnessing thrilling times in biological sciences. There is no limit to gather all the information from a living system—through the so-called *omics* techniques—but we lack the appropriate theoretical and conceptual tools to turn data into understanding. We are on the verge of synthesizing an artificial cell, but as John B.S. Haldane did anticipate, will this occur before we have a full understanding of those tiny chemical devices we call living cells? Some authors have suggested that synthetic biology is the armed arm of systems biology. Some others advocate for the straightforward application of engineering principles to life. We are convinced that through the synthesis of cells we will capture some essential aspects of life, let alone the immense horizon of technological applications forecasted for redesigned cells.

In this book, we will discuss on life, engineering and life engineering, and how to go beyond the boundaries of nature. We will describe the historical evolution of synthetic biology; from the term itself to the state of the art of the discipline; we will describe the complementary approaches to the ultimate goal of synthetic biologists: the creation of a truly artificial life form; we will make a focus on the iGEM competition; and finally, we will give an opinionated view on the boundaries of the discipline and its overlapping with other research fields, such as metabolic engineering.

Given the general interest in synthetic biology, but also the likely fear or euphoria associated with the possibility of an artificial production of life, we have worked on this book with the hope that scientists would find a non-biased guide to this emerging field, whereas an educated but non-specialist public may discover the clues to a better understanding of the actual scientific boundaries of synthetic biology.

Finally, we would like to recognize the help of several people and institutions in this work: Michel Morange and Ricard Solé, for kindly providing us with a Foreword and a Postface, respectively, with their authoritative perspectives from the present to the past and to the future; Fabiola Barraclough for expert proofreading of the manuscript. The financial support by Càtedra de Divulgació de la Ciència (University of València), Spanish Mineco (grant BFU2012-39816-C02-01), and Generalitat Valenciana (grant Prometeo 2009/092) is acknowledged. The Valencia-Biocampus iGEM team is supported by the University of València (Oficina de Polítiques d'Excel•lència) and the biotechnology company Biopolis SL. The work on this book has been supported in part by the European Union (grant ST-FLOW coordinated by Victor de Lorenzo).

Manuel Porcar
Juli Peretó

Contents

Chapter 1
What Is Synthetic Biology?

Abstract Synthetic biology aims at the design and construction of biological devices and systems for useful purposes. From an ideal engineering perspective synthetic biology works from rational design made through a few conceptual pillars, namely abstraction, standardization and modularity. Nevertheless, the combination of our still fragmentary biological knowledge and the messy nature of biological devices are major challenges for engineering life in a predictive manner. It is urgent to build bridges between different disciplines, from biology to engineer and back, to pursue this extraordinary goal of making life.

In an interview for a documentary film, the director of the Program on Emergent Technologies at MIT, Kenneth Oye, stressed: "the term Synthetic Biology seems to have been calculated to produce a negative reaction" (Schmidt and Meinhart 2009). True. Artificiality and nature do not combine well in most of our minds. And yet technology is one of the pillars of today's world and is without doubt one of the key factors behind the relative wealth and welfare of modern societies. Technology is also a powerful toolbox with which key global challenges such as climate change, food shortage or pollution issues can—hopefully—be tackled. Indeed, technology is both the cause of our historical success as a species and one of the best weapons for us to survive in the future.

In this book we will take a glance at one of the most promising new technologies: synthetic biology. But before we start, and in order to understand its very nature, we have to focus on its predecessor and sister discipline, biotechnology. In the vast range of relatively new technologies, modern biotechnology holds the world record for the fastest adoption. In less than two decades, transgenic crops, for example, have spread from their initial field tests prior to 1996 to their current overwhelming presence worldwide, with an enormous surface area dedicated to four main genetically modified crops today: soybean, cotton, corn and canola. Indeed, most of the cotton and corn on Earth are already transgenic cultivars.[1] But biotechnology is much more than transgenic plants. Genetically modified organisms produce drugs, synthesize biofuels, or carry out bioremediation. Biotechnology is

---

[1] International Service for the Acquisition of Agri-Biotech Applications (ISAAA). http://www. isaaa.org. Accessed 10 April 2014.

© The Author(s) 2014

M. Porcar and J. Peretó, *Synthetic Biology*,
SpringerBriefs in Biochemistry and Molecular Biology,
DOI 10.1007/978-94-017-9382-7_1

defined by the UN Convention on Biological Diversity as "any technological application that uses biological systems, living organisms or derivatives thereof, to make or modify products or processes for specific use". This definition encompasses many applications that one would not consider biotechnological at first sight: baking, brewing and farming—including animal domestication and plant cultivation. All these are biotechnology because they use living organisms for a practical application (i.e., to make bread, beer or anything edible). In a strict sense, though, modern biotechnology is restricted to direct genetic modification of any organism for a practical purpose. Sometimes, but not always, genetic engineering is used synonymously with biotechnology. Bioprocess engineering, metabolic engineering or bioengineering are branches of biotechnology which are often perceived as a proxy of biotechnology itself.

1.1 Engineering Ideals and Synthetic Life

But if all these branches of modern biotechnology deal with engineering organisms, what about synthetic biology? The term "synthetic" (man-made) immediately suggests the antinomy of "natural". But synthesis (from the ancient Greek σύνθεσις, σύν "with" and θέσις "placing"; "placing together") is originally defined not by human manufacture but as the combination of two or more parts, either by design or by natural processes. Synthetic biology is thus an ambiguous term, which might be taken as either "artificial biology" or as "constructive biology"; there is certainly a little bit of both in the discipline. The most accepted definition of synthetic biology is the design and construction of biological devices and systems for useful purposes. This definition is very similar to that of modern biotechnology but, interestingly, terms such as "design", "construction", "devices", and "systems" call to mind the central feature that places synthetic biology somewhat apart from biotechnology: its combined—blended—nature between biology and technology (Fig. 1.1). The idea is as simple as it is appealing. Since there are no doubts about the intrinsic power of technological and engineering developments (bridges stand, cars run, pencils write, internet works and so on) such an engineering approach to molecular biology should make life easier to engineer compared to usual biotechnological approaches.

It has to be stressed here that biotechnology has been developed mainly by biologists, who, as is often the case in this classical discipline, deal with an almost infinite range of (biological) variants. In biotechnology, molecular cloning largely depends on trial and error, and biotechnologists—maybe imbued with the evolutionary dogma of the central role of selection—tend to make chimeric DNA on an experimental basis: if a DNA construction works, it is kept; otherwise, it is simply assumed that other variants have to be tested. But synthetic biology is not inspired by evolution but by engineering, and engineering works from rational design made through a few conceptual pillars, namely abstraction, standardization and modularity.

Abstraction stresses the need to streamline many processes (life is an extreme example of this). The idea begins by defining abstraction levels that do not need to

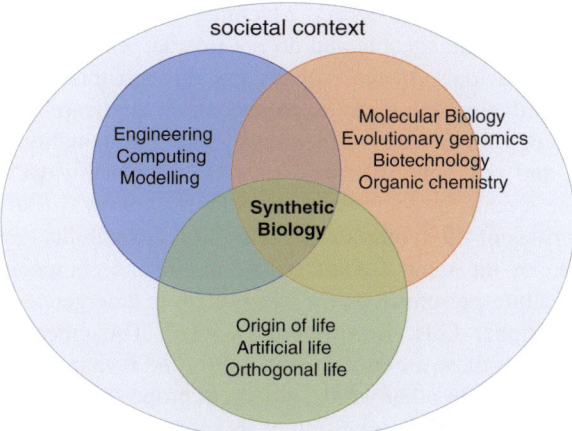

Fig. 1.1 Synthetic biology and its allied disciplines. Adapted from the European Science Foundation "Engineering Complex Biological Systems EuroSYNBIO" programme (Website accessed April 10th 2014 http://www.esf.org/index.php?id=5457)

be re-analyzed in every step of the construction process. Just as electric components are not demounted and tested individually but treated like black boxes (with the focus on the input/output rather on the particular internal workings), DNA cloning is proposed to make abstraction of the complexity (such as the DNA sequence itself; but also secondary structures, modulator interactions, etc.). With this approach, long complex sequences (ATGCTAGG…etc.) magically become biological parts, building blocks with which more complex associations (devices, circuits, etc.) can be mounted by the simple combination of individual parts.

These blocks, as in industrial engineering, are assumed standard. For instance, promoter sequences, i.e., pieces of DNA that modulate transcription initiation, can be scaled in a discrete range of strengths, like nuts and bolts. Thus, ideally, a synthetic biologist should be able to choose between a promoter of strength 5, 6 or 7, depending on the desired level of transcription. Promoters, reporter genes or pigment-coding sequences can thus be conceived as a palette (of colours, literally, in the case of pigments, or of strengths in the case of promoters) of defined strength/length/intensity/behaviour.

The combination of biological parts (a promoter of strength 7 and a fluorescent protein of strength 9, for example) might easily lead to devices or modules, which, in turn, are assumed standard and independent. Biological parts can be used as modules, that is, exchangeable blocks of defined function and useable in a range of microbial hosts and contexts. Modularity in biology is somehow different from the concept in industrial design, where modularity refers to the engineering technique that builds larger, more complex systems by combining smaller subsystems.

The concept of modularity is linked to orthogonality (etymologically, "straight angle"), which means independence. Two modules are orthogonal if they do not interact outside a defined interphase or, at least, do not interfere with each other.

This concept is central in engineering. Machines are designed and assembled in blocks, and these are exchangeable and do not interact with each other so as not to alter the overall behaviour. As we will discuss through this book, there is some debate among synthetic biologists as to the strict applicability of engineering constraints to biology. In the case of modularity and orthogonality, and considering the complexity and overlapping nature of metabolic pathways, could "relative orthogonality" be a realistic goal when engineering biology?

The particularities of life (complexity, flexibility, adaptability) represent both an obstacle and one of the main advantages of organisms compared to man-made machines. One feature present in every living form is emergence. Emergence was defined by philosopher G.H. Lewes, who wrote: "The emergent is unlike its components insofar as these are incommensurable, and it cannot be reduced to their sum or their difference". In other words, emergent properties are those that are not a direct outcome of the parts themselves; they are more than the sum of the parts. And this is something in conflict with engineering ideals. There is a famous quotation on emergence of one of the fathers of synthetic biology, Drew Endy. In an interview published in Edge,[2] he stressed: "I hate emergent properties. I like simplicity. I don't want the plane I take tomorrow to have some emergent property while it's flying." This is a very clear example of how biological complexity is a challenge for living beings to be engineered.

1.2 Challenges for Synthetic Life

Emergence is not the only challenge facing synthetic biology. There are both technical and organizational issues to be solved to firmly establish synthetic biology as the paradigm shift it is purported to be. In a review entitled "The ten grand challenges of synthetic biology" that one of us (MP) coauthored with other European researchers on the discipline, we defined a list of such challenges. These are logistic, technical and social (Table 1.1).

The first and the last challenges are social. The first requires the scientific community to reach a consensus on synthetic and streamlined genomes. It might not be obvious, but there are still difficulties in combining lexical and conceptual approaches from biologists and engineers and a common basis is required for the development of the discipline. The last challenge refers to science communication. After the disastrous Genetically Modified Organisms (GMO) battle, transparency should be one of the pillars of communication of synthetic biology if we want the discipline to be acceptable to the public. The remaining challenges are technical, and include difficult tasks such as modelling the complex and overlapping living circuitry or the challenge of combining engineering and selection-based approaches in the synthetic biology toolbox.

---

[2] http://edge.org/3rd_culture/endy08/endy08_index.html accessed April 10th, 2014.

Table 1.1 Challenges synthetic biologists have to deal with (from Porcar et al. 2011)

The ten grand challenges of synthetic life
1. Reaching a consensus on synthetic and streamlined genomes
2. Cooking from scratch (bottom-up)
3. Learning from nature: naturally evolved reduced genomes
4. Refine and make reality the notion of biological chassis
5. Manufacturing engineered biosystems
6. Overcoming physical and chemical constraints
7. From models to cells and back
8. Replication and reproduction
9. Towards an integrated design strategy of synthetic organisms
10. Coupling scientific development and public opinion information

Above all the challenges, there is the necessity to build bridges across disciplines (Delgado and Porcar 2013; Anonymous 2014). It is interesting to note that different visions coexist in synthetic biology today. Thus, there is no a unique synthetic biology approach, even less a unique definition or disciplinary borders. The different background of synthetic biologists mentioned above is at the basis of the controversial perception of the discipline, which is seen very differently from biological or engineering perspectives. Additionally, the first very ambitious predictions made on the success and applications of synthetic biology have been substituted by more realistic views, which are aware not only of the technical developments and milestones allowing more sophisticated, fast and relatively cheap genome modification, but also on the limitations of engineering life. There is still a controversy between the very conception of a living cell, which might be seen as a pure biological machine or as a radically different entity, arising from non-design and infested with emergent properties.

Background noise is a hot topic in many synthetic biologists' discussions. In engineering, a deviation of 0 and 1 might be due to noise in a strict sense, consequence of measurement bias, or correspond to loss of signal, such as in Internet data transfer. Often, values close to zero under a certain levels are corrected to zero; whereas values above a certain threshold are processed as ones. Background noise, interferences, signal loss and other factors mean that, even in the truly digital world, one and zero are not always pure values. In synthetic biology, promoter strength, fluorescent protein intensity, transcription force, enzyme activity and specificity, and many other factors have been quantified and scaled and assumed to behave in a scale from 0 to 1 or to 100. Compared to electronics or digital data transfer, biological processes are prone to avoid extreme values. Gene expression may be silenced, but there is often a certain leak. Biological noise—or messiness—is inherently associated to the nature of biochemical interactions, confined to soft chemistry in a nanometric scale. This important noise effect does not prevent biological circuits from working nor does it make biology impossible to engineer, but complicates the task of coupling and standardizing synthetic biological circuits enormously.

Mutation is one of the main features of life, and one of those that somehow sets biology apart from industrial production. Indeed, living creatures undergo various rates of random informational changes, which might have dramatic effects on adaptation, survival and evolution. Machines are designed to be more robust than flexible, and thus mutation-like processes do not usually occur. Interestingly, though, computer viruses, a source of "genotypic" (in computer science, hard drive) and "phenotypic" (computer behaviour) variation, work and spread very similarly to true viruses. Mutations, infections or—simply—time, always lead to the death of organisms. Machines are not immortal, but maybe it is farfetched to equate machine dysfunction with cell death because, for one thing, biological death is, by definition, irreversible, whereas machines can be repaired—by an external agent.

Extra-unit changes refer to the variation in the environment of a machine/cell. There are interesting similarities and differences between the relationship machines or cells have with their immediate environments. The former include the dependence of behaviour: for example many machines and bacteria work better at certain temperature ranges. Temperature, humidity, magnetic/trophic interferences, etc., are important factors in both biological and engineered systems.

It is certainly true that cells do not need to be biological machines to be genetically engineered. But it is also true that if a living cell is a very particular kind of machine, unveiling its complexity would rapidly lead to a fast and easy modification through standard engineering protocols. If one compares man-made machines and naturally occurring cells, several fundamental aspects are in contrast, and these include, at least: background noise and messiness, mutations and variability, unavoidable death, extra-unit changes, complex interactions with other units in a soft chemistry environment, a historical past (or phylogenetic dependency), and, last but not least, a particular internal organization that includes self-maintenance and self-repair, embodied in the fundamental notion of recursivity (see Chap. 3 for further discussion on cell-machine comparisons).

Machines are designed to fit and to stand with environmental conditions. In general they are made to work the same way on a very wide range of conditions. Cells, by contrast, flow with the environment and dramatically change their behaviour depending on that environment. Among the environmental factors with which interactions may occur, one of the most obvious is another unit (cell/ machine). Sex is a revolutionary tool for evolution and adaptation, but undesired informational exchanges between computers, for example, are to be feared. The spread of computer virus above mentioned is in part due to a lack of prophylactic attitude when exchanging informational portable devices such as USB keys. The similarity of computer virus spreading with that of sexually transmitted diseases (STD) is striking, and it is certainly due to the fact that both processes share fitness selection as a blind but yet driving force.

Another issue that is troublesome is phylogenetic dependency. When the first modern biotechnologists attempted plant transformation, they had to modify and adapt bacterial or viral DNA sequences to make them "suitable" for plants. In fact, this phylogenetic dependency means a lack of standardization, and it has been compared—again—with computer operative systems. But a problem in the metaphor is the number of such "operative systems" biological engineers have to deal with: even strains of subspecies might need important optimization and are recalcitrant to standard units (such as plasmids, for example). And, finally, a definition of the nature of life and the reason why the machine metaphor should be discarded: internal organization (recursivity, see Chap. 3). Cells exist, but since they have not been designed, they are built in a very different way compared to what an engineer would have done. Overlapping circuitry, complexity of interactions, emergent properties and a tendency to "do a lot with very little", or tinkering, clearly sets metabolic pathways apart from an engineering view of standard modules.

In summary, synthetic biology elicits diverse fundamental notions on living things and the possibility to engineer life. It is worth to mention that some practitioners, especially from the chemical field and referring to the history of their discipline, insist that synthesis is a research strategy and not a field, a strategy that enables us to explore problems, unveil discoveries and build new concepts in ways that observation and analysis cannot (see Chaps. 3 and 4).

References

Anonymous (2014) Beyond divisions. Editorial to a special issue on synthetic biology. Nature 509:151

Delgado A, Porcar M (2013) Designing *de novo*: interdisciplinary debates in synthetic biology. Syst Synth Biol 7:41–50

Porcar M, Danchin A, de Lorenzo V, Dos Santos VA, Krasnogor N, Rasmussen S, Moya A (2011) The ten grand challenges of synthetic life. Syst Synth Biol 5:1–9

Schmidt M, Meinhart C (2009) SYNBIOSAFE: Safety and Ethical Aspects of Synthetic Biology (DVD) Documentary Film, 38 minutes plus bonus material

Chapter 2
What Was Synthetic Biology?

Abstract The desire to make life is not new. Mythology and history provide numerous examples of this Promethean longing. Materialist and evolutionist scientists over a century ago were convinced of the possibility and even the need to synthesize living beings to advance the knowledge on the nature and origin of life. The premature synthetic biology attempts by Stéphane Leduc and Alfonso L. Herrera reflected the mechanistic ideal in biology of Jacques Loeb. The book "La biologie synthétique" by Leduc (1912) clearly defines the efforts of these pioneers: "Why is it less acceptable to seek how to make a cell than how to make a molecule?" Journalists have presented many advances in biology in the past century as an attempted synthesis of life. Nor is it new, therefore, the fine line which separates the scientific enthusiasm from hype.

The twentieth century may have witnessed the expansion and consolidation of biology in its myriad fields and many levels, ranging from molecular biology to ecology; however, the fundamentals of biological science date back to the previous century. It was in the nineteenth century that cell theory was developed, and Gregor Mendel, Louis Pasteur and Eduard Buchner performed their experiments; evolutionary theory was also put forward by Charles Darwin in the 1800s. In short, the nineteenth century marks the beginning of the materialistic study of living things. Thenceforth, life gradually broke away from the supernatural explanations that had escaped the realms ruled by the laws of physics and chemistry and lay beyond the bounds of the scientific method. The aspirations and intentions of those late nineteenth and early twentieth-century scientists were based on the belief that life could only have material explanations, and that they would only be able to understand it if they managed to make life in the laboratory.

© The Author(s) 2014
M. Porcar and J. Peretó, *Synthetic Biology*,
SpringerBriefs in Biochemistry and Molecular Biology,
DOI 10.1007/978-94-017-9382-7_2

2.1 Life and Matter

In 1868 Thomas Henry Huxley, personal friend and public champion of Darwin, gave a lecture entitled *On the physical basis of life*. Therein, Huxley referred to "protoplasm", a proteinaceous material harbouring all the properties of living things, as an object of study *par excellence*, the true physical basis of life which could be studied thanks to advances in "molecular physics". "[...] what community of form, or structure, is there between the animalcule and the whale; or between the fungus and the fig-tree? And, *a fortiori*, between all four? [...] if we regard substance, or material composition, what hidden bond connects the flower which a girl wears in her hair and the blood which courses through her youthful veins?" Huxley supported a view that not everyone shared at the time, namely that life was inseparably linked to matter and subject to physical laws and, although it might be at odds with common sense, "the physical basis or the matter of life was what united all living beings", namely, that sort of proteinaceous matter that was common to them all: protoplasm. Moreover, the properties of protoplasm would be the "product of a certain disposition of material molecules."

Then, in his 1870 speech as president of the British Association for the Advancement of Science, entitled *"Biogenesis and abiogenesis"* Huxley took a stronger intellectual stand. He acknowledged that life may have originated in the past from natural causes and did not rule out the possibility that life could be reproduced in the future, if the conditions enabling matter to acquire "vital" properties could be artificially established. Forty two years later, the president of the same Association, physiologist Sir Edward A. Schäfer, was to proclaim that the boundary between living and nonliving matter was so hazy that the only way to study the life phenomenon would be "by the same methods as all other phenomena of matter, and the general results of such investigations tend to show that living beings are governed by laws identical with those which govern inanimate matter". Schäfer entertained the idea of synthesizing some of the major components of the cell, Miescher's nuclein (our nucleic acids) and proteins, and did so with a very simplistic view built on the knowledge of cell chemistry in his day, which was somewhat inconsistent at the time. Schäfer finally stated his belief that life would be created in the laboratory: "The elements composing living substance are few in number [...]. The combination of these elements into a colloidal compound represents the chemical basis of life; and when the chemist succeeds in building up this compound it will without doubt be found to exhibit the phenomena which we are in the habit of associating with the term 'life' [...] The above considerations seem to point to the conclusion that the possibility of the production of life—i.e., of living material—is not so remote as has been generally assumed". Thus, the debate at that time was no longer whether life could be synthesized or not, but rather *when* this scientific breakthrough would be made.

Jacques Loeb provides the best example of the mechanistic ideal and the experimentalist endeavour in biology (Pauly 1987). Before moving to the USA, Loeb had worked with some of the most advanced scientists of his time in both physiology and chemistry: Adolph Fick, Julius Sachs and Svante Arrhenius. In 1896 Loeb explained his plans to set up a physiology laboratory at the University of Chicago. The laboratory—he said—would transform our view of nature and provide the following "services" to medicine: deal with the problem of famine, experimentally test the Darwinian explanation of the transformation of species, and "the most fundamental task of Physiology" was "whether or not we shall be able to produce living matter artificially". Thus, according to Loeb, they could not only demonstrate the validity of physiologists' ideas about biological phenomena, but also confirm the insignificance of beliefs in supernatural phenomena and, in doing so, convince the public they should trust in experimental scientists to direct social change. His positivist optimism led him to believe that science held the keys to progress, which could touch on all areas of human activity; a progressive science with unlimited prospects, even able to make life itself.

In 1906 Loeb published *The dynamics of living matter* which concluded by considering what the author deemed to be the two main issues facing biology: how to transform inert matter into living matter and how to transform a plant or animal species into another species. Loeb believed that the time had come to tackle these issues empirically and try to solve them. He was in favour of Pasteur's germ theory, rejected spontaneous generation (or heterogenesis) and criticized its proponents. Likewise, he did not accept that synthesizing proteins was equivalent to creating life or obtaining life-like forms. Thus he stated plainly that the aim to synthesize life would not be achieved by simply obtaining the substance of living beings (albuminoids, colloids…) but by obtaining a mixture of these substances that would possess life-like characteristics (self-preservation, growth and reproduction). He said the outer shape was secondary, thereby distancing himself from some experimental attempts to synthesize life, which we will refer to later. In his work of 1912, eloquently entitled *The mechanistic conception of life* (Fig. 2.1) Loeb set out to analyze life from a strictly physicochemical view and stated that "we must either succeed in producing living matter artificially, or we must find the reasons why this is impossible".

Loeb was what you might call a "visible scientist" in terms of media impact, but curiously he never actually tried to make living matter in the laboratory. However, he did become greatly renowned for achieving artificial parthenogenesis. In fact, he managed to make unfertilized sea-urchin eggs develop by simply changing the chemical composition of the surrounding medium. He was, therefore, able to replace the sperm with a chemical agent, which was taken to mean that biological processes must have a purely chemical basis. However, journalists at the time reported these experiments to be a real chemical creation of life, and some young ladies stopped bathing in the sea in case they got pregnant. Indeed, there was so much media hype that he felt obliged to publish a short note in the journal *Science*, in which he warned: "In view of the fact that a number of daily papers have printed reports concerning alleged or real experiments of mine I wish to state: (1) That none of the statements printed in the newspapers have been authorized by me. (2) That whatever I may have to say about my work will be published in scientific journals."

Loeb continued his quantitative experimentation, and rounded off his vastly diverse (artificial parthenogenesis, ion transport across cell membranes, animal

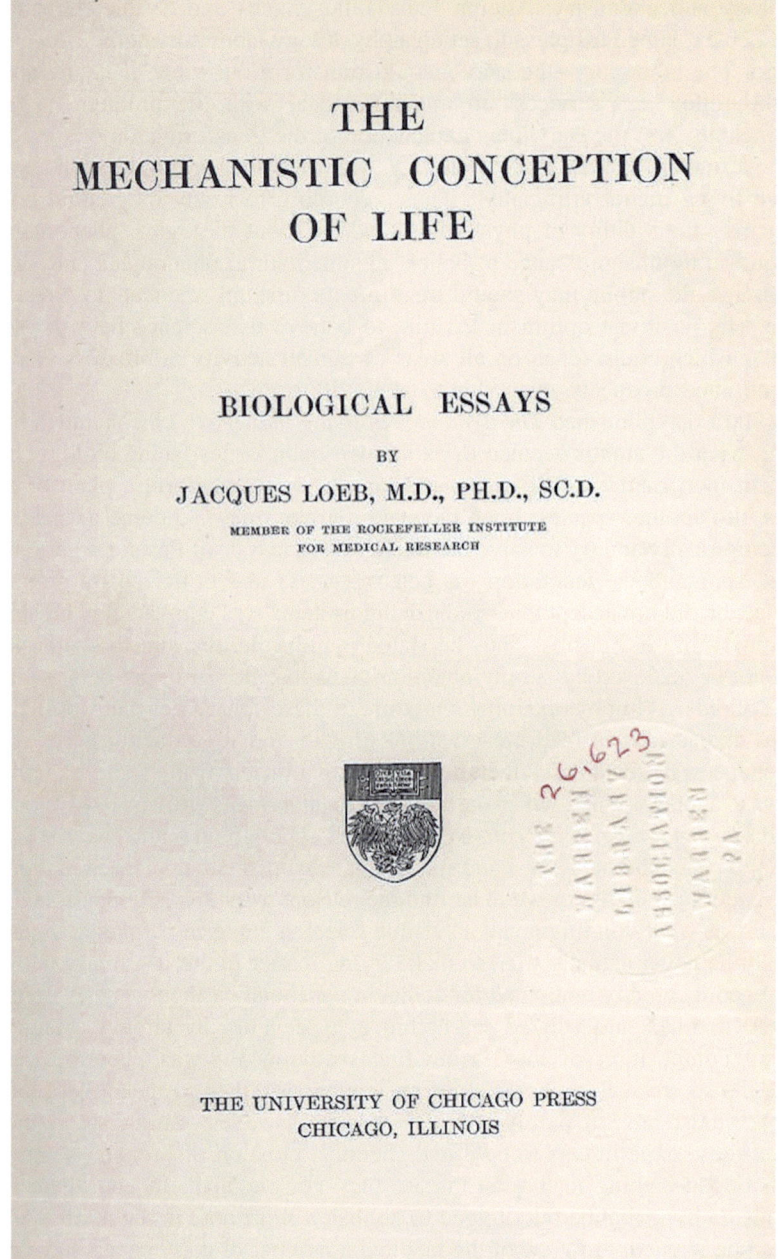

Fig. 2.1 Cover of Jacques Loeb's *The Mechanistic Conception of Life. Biological Essays* (1912)

tropisms) and distinguished career with a splendid study on proteins. His research helped to put an end to the idea held at the time that these cellular components did not obey the laws of chemistry. Thus, in the 1920s, scientists began to define the macromolecular nature of proteins and their physicochemical behaviour. Proteins were about to leave behind the domain of colloids, and biochemistry was to wake from what has been called "the dark age of biocoloidology". In the early twentieth century, the colloidal state—consisting of tiny molecules—was proposed to bridge the gap between chemistry and biology. However, new physical techniques made it possible to measure the size of biological molecules and to crystallize proteins, while structural methods based on X-ray diffraction revealed images of giant molecules (macromolecules), which could not possibly fit with colloids (Morange 2003). Discoveries about the chemistry of biological macromolecules also contributed to pointing scientists away from seeking the essence of life in its form rather than in its chemical composition, and to contemplate life as a result of the crystallization of inanimate materials, as we shall see later.

2.2 Spontaneous Generation

Even before Jean-Baptiste Lamarck—forerunner of evolutionary ideas—, spontaneous generation was recognized as an additional reproductive mechanism to sexual reproduction. Aristotle described it for many plants and animals, and even in Shakespeare's Mark Antony (*Antony and Cleopatra*) the mud of the Nile breeds life by action of the Sun. Many medieval legends believed in spontaneous generation and the existence of the so-called "goose-tree", which gave rise to fish or birds depending on whether its seeds fell into the water or onto land, causing Pope Innocent III to explicitly prohibit the consumption of geese and ducks during Lent given the popular belief of their origin. Francesco Redi's experiments in the seventeenth century, and the eighteenth-century controversy between Needham and Spallanzani, demonstrated spontaneous generation of animals to be impossible and shifted the belief in this process to the newly discovered microscopic world.

The pre-Darwinian view of nature was that all living beings were placed on the infinite rungs of a ladder leading up to heaven: The Great Chain of Being. The lower rungs held the minerals, progressing up through plants and humbler animals, such as worms, to the penultimate step where man was placed, preceded only by the angels on the stairway to God's kingdom (Fig. 2.2). Carl Linnaeus, who introduced the current system of binomial nomenclature, devoted his life to putting each and every living being in its place, each one shaped and designed by the Creator. Then in the early nineteenth century, Lamarck began to shake the ladder, and thus worms might become men. Every living being was driven towards achieving perfection, which, together with the use or disuse of organs, made organisms evolve. Through evolution, living beings on the lower rungs could move a step up. But where did the beings on the first step come from? Following the noble precedent of the Count of Buffon, Lamarck proposed that the simplest life forms, at the base of the ladder,

Fig. 2.2 The great chain of Being or the Lullian staircase. *De noua logica, de correllatiuis, necnon [et] de ascensu [et] descensu intellectus* by Ramon Llull (published in València, 1512 by J. Costilla). Reproduced with permission of Historical Library, University of València, ref. BH R-1/341(1)

appeared through spontaneous generation. Thus, for the first time spontaneous generation was envisaged not as an alternative reproductive mechanism to sexual reproduction, but as an explanation for the very origin of life.

In 1859, coinciding with the publication of Darwin's book *On the Origin of Species*, Félix Pouchet published his *Hétérogénie ou traité de la génération spontanée basée sur de nouvelles expériences*, an extensive treatise which claimed to demonstrate spontaneous generation through numerous experiments. Such was the impact of the work that the French Academy of Sciences convened an award for scientists to demonstrate the existence, or not, of spontaneous generation. Finally, in 1862, the prize was won by the already famous chemist Louis Pasteur, for a series of brilliant and immaculate experiments showing the mistakes made by Pouchet. Although—strictly speaking—Pasteur did not repeat Pouchet's experiments and, therefore one might argue it was not a rigorous scientific rebuttal, the methods and instruments devised by Pasteur have gone down in history as one of the most remarkable examples of scientific reasoning. Pasteur in France and, later, John Tyndall in Britain—confronting Henry C. Bastian—almost managed to dismantle the ancient belief in spontaneous generation with the help of experimental scientific methods. However, as we shall see later, several authors attempted to skirt around Pasteur and Tyndall's hurdles.

2.3 The Synthesis of Living Beings a Century Ago

The definition of synthetic biology in the late nineteenth and early twentieth century revolved around the idea of making living things from purely physical and chemical ingredients. This concept could be traced back to two authors of reference: France's Stéphane Leduc and Mexico's Alfonso L. Herrera (Peretó and Català 2007). Leduc was professor of biophysics at Nantes medical school and was considered to be the main exponent of synthetic biology in Europe. He gained popularity for his work on osmotic growths in his day and was an author of reference for D'Arcy Thompson; however, he left very few traces of his scientific activity behind. Although his efforts to synthesize life may seem absurd today, as the historian and philosopher of science Evelyn Fox Keller recognizes, they constitute "an episode in the history of biological explanation, the ambitions those efforts reflected, as well as the interest they evoked in their time (Fox Keller 2002)".

Leduc thought, as did Loeb and Herrera, that there was continuity between the inanimate world and living beings, and that an understanding of the underlying biological mechanisms could be gained through synthesis. The year 1901 marks one of Leduc's first publications, a communication at the Congress of the French Association for the Advancement of Science in Ajaccio, entitled *Cytogenèse expérimentale*. Herein Leduc described how to synthesize cells and concluded that these artificial cells were identical in shape to living cells, and had the same organs, nucleus, cytoplasm, envelope membranes, as well as their main functions, cell metabolism and evolutionary capacity. He claimed that his experiments refuted two doctrinal statements: the first proclaiming it impossible for living matter to be organized under the sole influence of physicochemical forces; and, the second stating that a cell cannot form spontaneously, and that every cell originates from a previous cell. Thus, in one foul blow, Leduc struck at vitalism and the impossibility of spontaneous generation, which were two of the cornerstones in the work of scientists like Pasteur.

After numerous communications to the Academy of Sciences, which sparked passionate debate among his colleagues and received some sharp and devastating criticism, Leduc undertook what would be his most outstanding work in the mechanistic study and experimental exploration of living matter. This was a series of three books entitled *Études de Biophysique* published between 1910 and 1921, of which the second volume was called *La biologie synthétique* (1912) (Fig. 2.3). This most probably marks the first time the term *Synthetic Biology* was used in a scientific work (Peretó and Català 2007; Campos 2009). Leduc's main proposal was that osmotic pressure was the only physical force required to generate amazing organic forms. The scope of synthetic biology ranged from the synthesis of organic molecules—fully consolidated through nineteenth century organic chemistry—including the synthesis of cells and tissues, to more complex structures. But Leduc wondered why organic synthesis was so well established and generally accepted while other stages were not only neglected but often treated with disdain. "Why is it less acceptable to seek how to make a cell than how to make a molecule?" Leduc wondered.

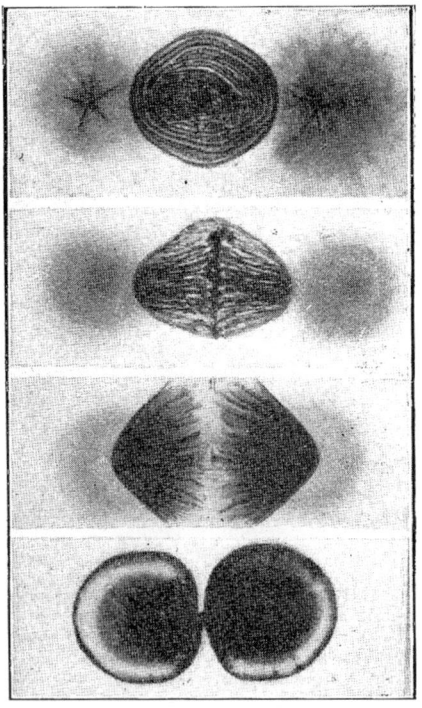

ÉTUDES DE BIOPHYSIQUE

LA BIOLOGIE SYNTHÉTIQUE

PAR

STÉPHANE LEDUC

PROFESSEUR A L'ÉCOLE DE MÉDECINE DE NANTES

AVEC 115 FIGURES DANS LE TEXTE

A. POINAT, ÉDITEUR

121, BOULEVARD SAINT-MICHEL ⚹ PARIS

1912

Fig. 2.3 *Left* Cover of Stéphane Leduc's *La biologie synthétique* (1912). *Right* Detail of figure on page 125 "four successive periods of karyokinetic division reproduced by diffusion"

This French scientist regretted the lack of attention paid to synthetic biology by academia, which considered his interpretations so fanciful that they could not possibly be taken seriously; however, his work gained great popularity with the lay public. The spectacular nature of osmotic growths, also known as chemical gardens, gained them access to a great many living rooms in the form of family entertainment. Leduc himself claimed that it was a wonderful sight to see a shapeless piece of calcium salt turn into a shell, coral or fungus, and all as a result of simple physicochemical forces. The anticipation and emotions stirred up by the fans of these chemical experiments have been immortalized in Thomas Mann's novel *Doctor Faustus*. The narrator describes the atmosphere created when onlookers contemplate the strange forms resulting from experiments made by Jonathan (father of the musician Adrian Leverkühn): " 'And even so they are dead', said Jonathan, and tears came in his eyes, while Adrian, as of course I saw, was shaken with suppressed laughter. For my part, I must leave it to the reader's judgment whether that sort of thing is matter for laughter or tears."

Leduc's work caused a stir even before his books were published. In 1902 the journal *The Academy and Literature* referred to a communication made by Leduc at the aforementioned Congress of Ajaccio, in which he explained his work of synthesizing cells from various chemicals. The commentator praised these efforts, linking them with speculation about the origin of life, adding that if they were well-founded, "the homunculus of Paracelsus, although it may never come to us in visible form, yet may not be such an impossible dream".

In all truth, the many illustrations showing the results of Leduc's experiments (like plant or fungal forms, cells dividing…) are of great beauty, and it is quite understandable that they should stir great public interest given their remarkable resemblance to living forms. Leduc's work is a nice example of how inorganic structures may strikingly resemble biological morphologies and textures (for further discussion on biomimetic materials, see below).

On the other side of the Atlantic, in Mexico, the prominent biologist Alfonso L. Herrera was a driving force behind several institutions introducing an evolutionary approach to the study of biology. Herrera is perhaps better known as the father of what he called *Plasmogenia*, a science aiming to synthesize life in the laboratory based on inorganic materials and which would unravel the enigma surrounding the origin of life, among other questions. His conviction that there was absolute continuity between inert matter and living matter is clearly expressed in what is considered the first Mexican biology text book: *Nociones de biología* (1904) "live pseudo-beings and pseudo-organized structures have been made in the laboratory, using reagents that are neither mysterious nor divine […]. Indeed, they are so suggestive of analogies between animate and inanimate matter that the spirit is confused […] and hesitates before drawing the final and definitive conclusion that *there is no separation between living forms and crystallized forms*". Two years later, his book was published in French under the title *Notions générales de biologie et plasmogénie comparées*.

Herrera published his unstinting work on Plasmogeny in the *Bulletin du Laboratoire de Plasmogenie* that he edited himself. In 1932 he contributed to a

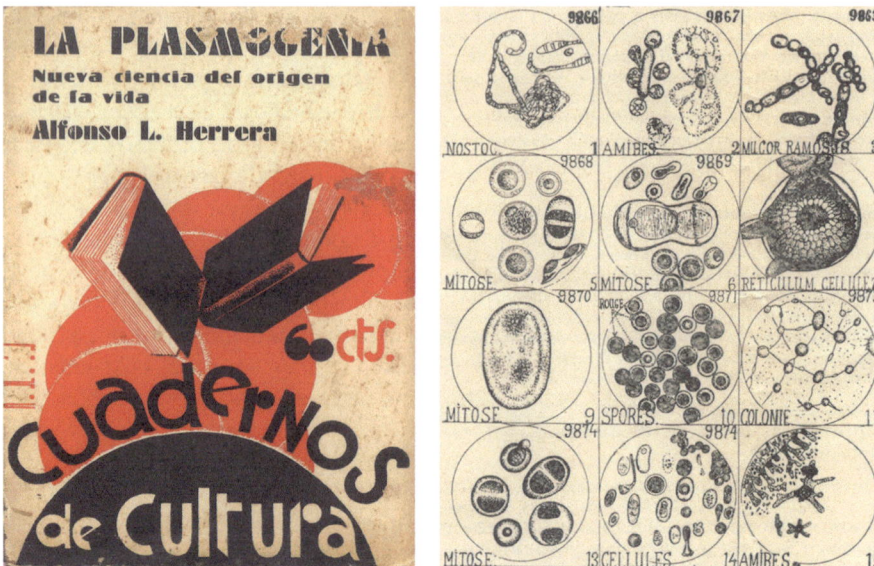

Fig. 2.4 *Left* Cover of *La Plasmogenia: Nueva ciencia del origen de la vida* by Alfonso L. Herrera, Valencia (1932), English translation available in Cleaves et al. (2014). *Right* Detail of the cover of *Bulletin du Laboratoire de Plasmogenie* 1(99) (1940)

collection of booklets published in Valencia, entitled *Cuadernos de cultura,* devoted to disseminating knowledge among the general public under Spanish II Republic cultural effervescence, with his monograph: *La plasmogenia: nueva ciencia del origen de la vida* (Plasmogeny: a new science on the origin of life) (Fig. 2.4). He clearly stated: "the problem of Plasmogeny is, simultaneously, morphological, concerning the imitation of forms; chemical, concerning the reproduction of elemental composition; and physical, concerning the reproduction of the physical conditions under which life is produced […] in particular with those assumed to exist in the earliest ages of the Earth". But Herrera thought the idea that the main chemical component of living matter was protein-like—an idea sparked, as we have seen, by Huxley's conception—was wrong; indeed, he focused his studies on minerals instead. Furthermore, the Mexican scientist condemned creationist religious prejudice, which clung to Pasteur's experiments as proof against spontaneous generation. For this reason, he stated that "the Church, worshipper of Pasteur, fanatical genius, alas, foams at the mouth with rage at Plasmogeny". And indeed, it was the Catholic scientists who hurled the sharpest spears at Herrera and Leduc's work: Jaume Pujiula, Spanish Jesuit and embryologist; Jean Maumus, French priest, physician and cell biologist; and the Italian, Agostino Gemelli, Franciscan, psychologist and biologist.

Although freer from religious prejudices, some biochemists like Jacques Loeb, criticized Leduc and Herrera's efforts for being premature or for straying far from the path of biochemical knowledge of their time. Undoubtedly, developments in biochemistry would temporarily move priorities away from scientists' dreams of synthesizing artificial life. For these critics, wishing to reduce life to its physico-chemical material basis may be legitimate; however, the way forward was not by using simple inorganic matter, as Leduc and Herrera did, but through colloids and proteins, which were beginning to reveal themselves to be giant molecules that were more difficult to study (Deichmann 2012). Therefore, studies should first endeavour to thoroughly characterize and seriously investigate the biochemical basis of life. Future advances were to show that not only should attention be paid to proteins—the true driving force of cell activity—but also to nucleic acids, which would take the driver's seat after the genetic studies of the 1940s. And later still, a third ingredient was to be added to attempts to reconstruct the simplest cell: the membranes formed by amphiphilic molecules.

Although we now know that the processes underlying the generation of those mineral, inorganic structures and life forms are quite different, the classic problem of studying the origins of natural structures through their morphological features is still unresolved. Indeed, establishing the biogenicity of structures—namely the chemical and/or morphological signature of past life—in the oldest geological records (i.e. microfossils and stromatolites) is a challenging and controversial topic. Likewise note that microscopic forms found in the Martian meteorite ALH84001 were interpreted by some as microfossils, an extraordinary claim that remains unproven. These are vivid examples of how this issue is still of great relevance to

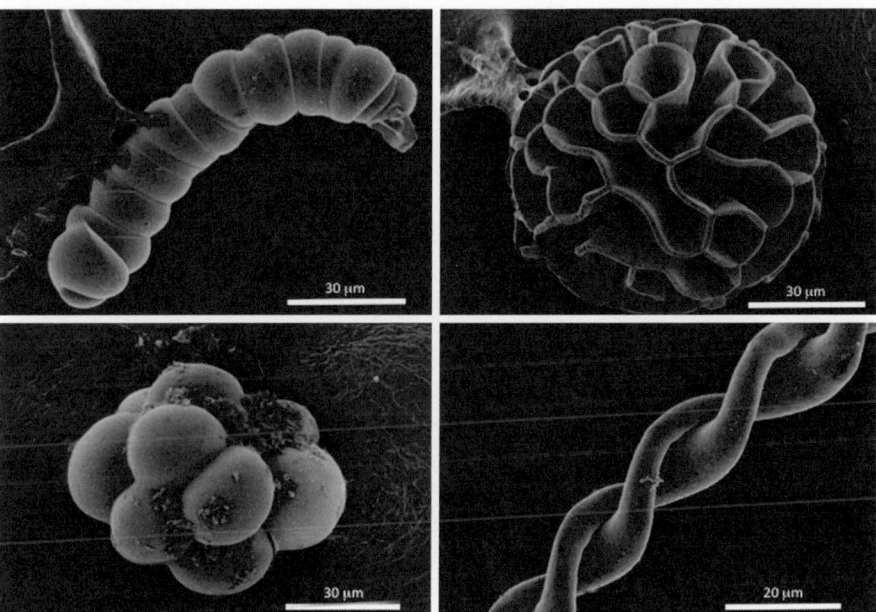

Fig. 2.5 Silica biomorphs. These microscopic objects self-organize by slow crystallization of barium carbonate in silica polymers at alkaline pH. Micrographs courtesy of Juan Manuel García-Ruiz (CSIC, Granada)

science today, namely, a key to dating of the origin of life and to identifying extraterrestrial life forms. The study of biomimetic chemical structures that self-organize under specific environmental conditions is an active research field in material science and biomineralization. Silica gardens (à la Leduc) and silica biomorphs (Fig. 2.5) are outstanding examples of these phenomena. Silica biomorphs offer a wonderful variety of elaborate morphologies with smooth curvature, very reminiscent of biological forms (Fig. 2.5).

2.4 Creating Life: Utopia and Propaganda

This idea of creating life artificially would resound among scientists and journalists throughout the twentieth century. The former perceived it as a far-off utopia, the latter as an achievement worthy of front-page headlines. So it was the Russian biochemist Aleksandr I. Oparin who linked the study of the origin of life with the experimental reproduction of the first steps in the evolutionary process. Oparin embodies the triumph over the scientific conflict and intellectual tension caused by accepting Darwinian evolution—with all its implications—and by acknowledging the irrefutable impossibility of spontaneous generation—as demonstrated by Pasteur and Tyndall. To be an unwavering Darwinian, meant admitting that all living things have a common origin and that, consistent with the rest of evolutionary theory, the origin of life could be traced back to purely material causes. According to Oparin, life originated as a result of a process of chemical evolution on a primitive Earth, where the right components, ingredients and physical conditions coincided, giving rise to the first elementary cells. In his first work, dating back to 1924, Oparin adopted an eclectic position that would allow biochemical innovations to be incorporated to his explanatory outline.

Although his 1924 pamphlet in Russian was not translated into English until 1967, his more comprehensive treatise published in 1936 took just 2 years to come out in English, and became widely known and readily accepted. This book begins with a rational argument against spontaneous generation and panspermia. The Russian author then went on to give a detailed explanation of the origin of life by chemical evolution. Oparin believed that it was legitimate to culminate experimental research on the origin of life by synthesizing life in the laboratory, stating: "the road ahead is hard and long but without leads to the ultimate knowledge of the nature of life. The artificial building or synthesis of living things is a very remote, but not unattainable goal along this road".

The work by Stanley L. Miller and Harold C. Urey, who managed to simulate the primitive prebiotic synthesis of organic molecules in 1953, was reported in the media as a step towards synthesizing life. At the University of Chicago, Miller and Urey had mixed gases that were thought to be components of the early atmosphere (hydrogen, methane, ammonia and water vapour) and subjected it to an electric current. After a few hours the condensate liquid—simulating primitive seas—changed colour, indicating it contained new substances. Subsequent analysis identified amino acids identical to those that make up proteins. This work had a huge media impact and even led to a Gallup poll, asking Americans if they thought it would be possible "to create life in a test tube". Only 9 % of respondents answered affirmatively while 78 % answered negatively and 13 % did not know what to think. Needless to say, there is a huge gap between the complex mixture of organic molecules and the simplest cell.

Despite the fact public opinion seemed little inclined to think it possible to artificially synthesize life—recognizing the enormous complexity and difficulty of the project rather than religious reasons—, journalists did not miss the next opportunity to bring Frankenstein's ghost back to the front pages. This came about in 1955, with the artificial reconstruction of tobacco mosaic virus (TMV) by Heinz Fraenkel-Conrat and Roblay Williams, at the University of California at Berkeley (Creager 2002). These researchers managed to obtain infectious TMV particles simply by mixing pure protein from the virus with genomic RNA, which were not infectious separately (although shortly afterwards, the RNA was shown to be pathogenic on its own). While there is an obvious connection with the artificial synthesis of life—given some researchers thought viruses may be the earliest primitive life forms—this biochemical achievement was to be expected. Indeed, it was a logical consequence of the self-assembly capacity of macromolecules and the biological role attributed to each of the viral components, and did not have any direct connection with scientific issues on the origin or synthesis of life. However, we could extend this debate to include other questions, such as: Is a virus a living thing? Does the self-assembly of macromolecular components taken from a pre-existing virus actually count as synthesis? In any event, the University of California managed to win the media attention they sought.

It would be worth studying this desire to present breakthroughs in biochemical research as the synthesis of life in the laboratory. A desire shared by journalists and politicians alike. In his autobiography *For the Love of Enzymes: The Odyssey of a Biochemist*, Arthur Kornberg (Kornberg 1989) recounts how a hundred newspaper and television reporters flocked to the press conference convened at Stanford University in 1967 to announce the enzymatic synthesis of the genome of the PhiX174 virus in a test tube, just one more chapter in a long series of studies into DNA synthesis by Kornberg. This was the first synthesis of a DNA viral genome that turned out to be biologically active. Kornberg himself warned the Stanford press office to avoid using phrases like "synthesis of life in the test tube" at all costs. Despite these precautions, worldwide the mass media made allusions to the creation of life in the laboratory. The same day, President Lyndon B. Johnson was taking part in an event to celebrate the *Encyclopaedia Britannica* bicentenary at the Smithsonian Institution in Washington and, ignoring the text provided by the Stanford press office, said "What are you going to read about tomorrow morning? It is going to be one of the most important stories that you ever read, your Daddy ever read, or your Grand-pappy ever read… Some geniuses at Stanford University have created life in the test tube!". Alistair Cooke hit the nail on the head, stating in the Manchester Guardian Weekly: "It is near enough to the truth to astound the layman, far enough away to annoy the expert".

In more recent times, similar media hype has surrounded the work of scientists led by J. Craig Venter (see Chap. 4). Not lacking pretension, Venter presented his experiments shrouded in the mystique of life created in the laboratory. Doubtless, he himself has helped to build an image of someone who denies playing at God while behaving likewise. It may be true that the synthesis of a complete genome and its successful transplantation in a cell is a technological breakthrough; however, it is debatable whether the resulting cell is entirely artificial or whether, in fact, it is a mere imitation, a simple copy of life as we know it.

References

Campos L (2009) That was the synthetic biology that was. In: Schmidt M et al (eds) Synthetic biology: the technoscience and its societal consequences. Springer, Dordrecht, Chap 2, pp 5–21

Cleaves HJ, Lazcano A, Ledesma Mateos I, Negrón-Mendoza A, Peretó J, Silva E (2014) Herrera's 'Plasmogenia' and other collected works: early writings on the experimental study of the origin of life. Springer, New York

Creager ANH (2002) The life of a virus. Tobacco mosaic virus as an experimental model, 1930–1965. The University of Chicago Press, Chicago

Deichmann U (2012) Crystals, colloids, or molecules?: Early controversies about the origin of life and synthetic life. Perspect Biol Med 55:521–542

Fox Keller E (2002) Making sense of life. Explaining biological development with models, metaphors and machines. Harvard University Press, Cambridge

Kornberg A (1989) For the love of enzymes the odyssey of a biochemist. Harvard University Press, Cambridge

Leduc S (1912) La biologie synthétique. A Poinat, Paris

Morange M (2003) Histoire de la biologie moléculaire. La Découverte, Paris

Pauly PJ (1987) Controlling life Jacques Loeb and the engineering ideal in biology. Oxford University Press, New York

Peretó J, Català J (2007) The renaissance of synthetic biology. Biol Theor 2:128–130

Chapter 3
What Is Life?

Abstract Despite the difficulty of defining a living being, biological sciences have considerably advanced. Today many authors feel the need to revisit the issue of the definition of life, among other reasons, because we are very close to have a second example of life. This life will not be the direct result of more than 3,500 million years of evolution, but the outcome of a project of synthetic biology in a laboratory. The fact that evolution has explored only a small part of the possible may pave the way towards alternative artificial lives. Focusing on the nature of life makes us more critical with the Cartesian comparisons between cells and machines. At the same time, progress in synthetic biology will allow us to better understand the organization of biological complexity.

It may sound paradoxical, but biologists have shown little interest in asking "what is life?" They have put a far greater energy into describing and classifying living things than to questioning the nature of life itself. Paging back through the history of the concept of life takes us back to Aristotle and beyond (Bedau and Cleland 2010), but the first scientist to ponder on "the essential conditions that were necessary to bring all the living species into existence" was Lamarck. He was one of the first to introduce the term biology and was convinced that all living beings could be accounted for by "purely physical phenomena". Lamarck was not satisfied with simply describing and studying the properties of living beings, he thought it necessary to reveal the essential characteristic that differentiates living beings from inorganic matter. For this reason, François Jacob argues that the 'concept of life' did not really exist before the nineteenth century.

By contrast, Charles Darwin saw no problem, stating: "It is no valid objection that science as yet throws no light on the far higher problem of the essence or origin of life...", and continued by saying that, similarly, not knowing the essence of gravitational attraction did not prevent physics from advancing. For Darwin, it was premature to ask what life was or how it originated (Peretó et al. 2009).

In more recent times, interest in the definition of life has been overshadowed for decades by the pursuit of molecular biology epitomizing the twentieth century. Life started to be depicted in molecular terms, and this seemed sufficient to advance our understanding. Indeed, according to James Watson, Francis Crick walked into the

© The Author(s) 2014
M. Porcar and J. Peretó, *Synthetic Biology*,
SpringerBriefs in Biochemistry and Molecular Biology,
DOI 10.1007/978-94-017-9382-7_3

Eagle pub, near the Cavendish Laboratory at the University of Cambridge, where they constructed the double helix model of DNA, and announced that they had 'found the secret of life'.

For molecular biologists, to wonder 'what is life' was a secondary issue, even though some of those responsible for these developments retrospectively recognized the influence of a booklet published by Erwin Schrödinger in 1944 entitled *What is life?* But Schrödinger had sought the answer in the wrong sphere (subatomic physics), ignoring the proper domain: chemistry.

Despite being eclipsed by molecular biology, this question was kept alive by the research programme on the origin of life, underway since Oparin's pioneering work started in the 1920s (Lazcano 2008). The Russian biochemist established a close connection between the definition of life and its evolutionary origin, since for him "an understanding of the nature of life is impossible without a knowledge of the history of its origin".

But can we understand life from its molecular description alone? The post-genomic era blatantly shows this is not so, however many genomes we may accumulate in databases, our understanding of life barely progresses: a surfeit of information but a scarcity of knowledge. Indeed, the bounds are no longer shaped by the technology necessary to collect and accumulate data, but rather by our ability to untangle so much data and reach meaningful conclusions. How can we overcome this overwhelming mismatch?

As Michel Morange has pointed out, there is currently a renewed interest in understanding the essence of living beings and how to apply these advanced theoretical approaches to this conundrum (Morange 2003, 2012b). These novel approaches include: (a) the growing interest in systems biology—a way of *thinking* about biological systems and making sense of the mountains of genomic or other kinds of data—or what is called astro- or exobiology; (b) the popularization of infra-biological models such as the RNA world hypothesis on the early stages of cellular evolution; (c) the development of computational approaches to life (the so-called artificial life or *ALife*); or (d) the enthusiasm sparked by synthetic biology around the possibility of synthesizing life in the lab in the near future.

3.1 Why Wonder What Life Is?

Jacob acknowledged that no one in the lab asks what life is. We know what we know about the cell and how it operates without having a consensus on the definition of a living being, one that establishes the necessary and sufficient conditions for a piece of matter to be considered alive. We can investigate and examine the composition, structure, transformation and perpetuation of living matter without having to define it. This has been, in short, the real history of biology. There is also the view that life cannot be defined unless we have a general theory of living beings. We must understand life in depth before trying to define it.

One way to focus the extensive literature on the definition of life (Bedau and Cleland 2010)—and, above all, the wide array of formulations—is to consider what kind of response you want. You may be interested in a descriptive definition, in order to propose defining criteria or diagnostic methods to classify an object as being alive. Such is the case of exobiology or synthetic biology: identifying a second example of life found on another planet or made in a laboratory.

For example, how can we design methods to detect extraterrestrial life? How can we agree whether or not life has been found on another planet? What criteria can we use to determine whether a piece of matter is alive or not? There is no doubt that technological and scientific advances enable us to explore our cosmic environs directly or indirectly. Thus, through remote analysis or by sending probes, we will gain greater knowledge of the solar system, and especially of those objects that may harbour life or their fossilized remains.

The quest to make life in the lab is another obvious case calling for agreement on the definition of a living being. Therefore, if someone undertakes a research project aiming to synthesize a living being from simpler components, how do we know whether they have actually achieved their goal if we do not agree on *what* should be obtained? Do we settle for a molecule that can copy itself, even though it is with the researcher's help? Or perhaps something that looks like a cell, able to nourish itself and reproduce?

Therefore, if we mean to venture further in the laboratory or cosmos, we should first discuss definitions (Ruiz-Mirazo et al. 2004, 2010). Here we adopt the eclectic position that, given current biological knowledge, it is possible to test a definition, which would serve as a starting point for discussion and comparison with other proposals.

3.2 A Single Example of Life

The first difficulty we face is that we only know one kind of life: life on Earth. Nonetheless, by observing and studying known life forms, we can draw up a list of properties and characteristics shared by all living things and try to interpret this universality in genealogical terms. Universality obviously refers to the entire diversity of life on Earth, and summarizes those components and processes inherited during evolution. Indeed the principle of evolution—of certain beings descending from others throughout the planet's history—is the most solid, efficient and simple explanation of this universality. Darwin surmised that all terrestrial organisms have a common origin and biological science—accrued over the last one hundred and fifty years—has confirmed this. With the universal list of shared biological traits we could test its essence and compression in one preferably brief sentence Ruiz-Mirazo et al. (2004). This could be one strategy. Harold J. Morowitz (1992) has comparatively studied the characteristics of extant living matter on Earth and has proposed a list of fundamental generalisations. The cellular nature of life is a universal property with at least two fundamental corollaries: (a) cell theory states that *every cell is derived from another cell*, expressing the continuity of epigenetic inheritance; besides the genetic material, cellular components, membranes, and other structures are inherited during reproduction that cannot appear de novo; (b) The cell is the smallest unit of energy transformation, cells are dissipative systems and flows of external energy and matter are channelled by compartment (or cell)-dependent chemiosmotic mechanisms. These two observations on terrestrial life are linked to two fundamental pillars upon which the living phenomenon—as we know it—rests: structure (reflected here by a semi-permeable membrane separating the inside from the outside) and the dynamic organization of matter (in this case cell-based bioenergetics (Ruiz-Mirazo et al. 2004)).

3.3 Real Is a Small Part of the Possible

The omnipresence of carbon compounds in cells illustrates the fact that we live in an essentially organic universe. This is what is observed from Earth: the most abundant elements in the universe (except helium, a noble gas) are hydrogen (H), carbon (C), oxygen (O) and nitrogen (N). Carbon chemistry can be seen in every nook of the observable universe. Therefore, the chemistry upon which life is based is the most common chemistry of the universe, an extension of the Copernican principle of mediocrity. Life is based on common chemistry, on an ordinary planet orbiting a commonplace star.

The specific organic molecules forming terrestrial life are a small subset of all the potential ones, and are most likely the result of historical contingencies, evolutionary decisions made once and for all, with no turning back. This is an excellent illustration of Jacques Monod's idea that life is the result of combining chance and necessity, with the two sides of the coin being what cannot be otherwise in the material universe we inhabit, and what various potential solutions would allow, which are all more or less equivalent. These decisions may be the result of an optimization process, of choosing the best solution among those possible, or may be the reflection of a frozen accident, a molecular fossil of no return. For example, the four nitrogenous bases (adenine, guanine, cytosine and thymine) constituting current DNA could be the result of a fortuitous choice very early in evolution. There are other similar compounds that could perform their function equally well. Similarly we can ask why our DNA is made up of four different bases rather than two or six. Eörs Szathmáry has concluded that two were too few and six too many for the molecular characteristics of the most primitive organisms and, therefore, four proved an optimum number, a *decision* frozen in time, unchangeable because evolution can never be rewound. Some chemists, Albert Eschenmoser and Ram Krishnamurthy among them, have explored the properties of other possible biochemical structures (*life as it could be*). Perhaps it is necessary to have a non-monotonous polymer as a structural support of genetic information, possibly the replication mechanism based on molecular recognition, a specific fit, is also a must. But this can be achieved with many different molecules.

The same can be said of metabolic pathways: why do certain transformations occur and not others? Somehow, the selection of certain types of organic molecules defines what kinds of changes can occur between them. Evolutionary success involves selecting between potential multiple reactions (thermodynamically), a kinetically feasible subset, i.e., occurring in reasonable time scales, at the mercy of the participation of catalysts: enzymes. These not only ensure reaction type and molecular specificity but also restrict the range of possible products.

The intrinsic flexibility of proteins means that the main enzyme activities are accompanied by other secondary activities, so-called *enzymatic promiscuity*. Natural selection operates on a primary function, but there may be other minor functions that are not subject to selection, and which enable the system to adapt to environmental change and evolve (i.e., evolvability of the system) (Tawfik 2010). The evolutionary trajectories of proteins undergoing structural changes have been studied experimentally to investigate how certain functions are committed to structural stability. There is a real trade-off between the various functions that allows for the evolutionary domain to be entered and new adaptations explored (Tawfik 2010). However, the actual raw material for evolution in biological systems can also be a drawback to its engineering and redesign. We would expect biological components to have a predefined function; likewise accessory functions that appear unexpectedly or out of control would not be desirable for the engineer. Perhaps enzyme promiscuity is one of the major challenges that synthetic biology must face, particularly in the development of orthogonal systems, namely, where the cross-talk between cell components should be minimal.

But what differentiates living matter from complex organic chemistry in liquid water? Certainly, the most striking thing about the chemistry of life is the *way* it occurs, how it is organized, rather than the specific molecules involved. So, to start with, the inside of the cell is crowded with matter and the few gaps therein are filled with water. In other words, it is a thick chemistry. But the real transformation of chemical bonds does not occur in the thick soup, or in the water. Instead, all the reactions are catalyzed by enzymes, three-dimensional structures that are soluble or wrapped up in membranes, containing a rough anhydrous hollow, the catalytic site, where the miracle of specificity and catalytic proficiency takes place. It is important to remark that the behaviour at these nanometric scales is not comparable to the performance of human scale systems (e.g. electronic and mechanical devices). This is one of the major difficulties when comparing cells and human-designed machines.

Biochemical processes operate far from equilibrium, leading to what Addy Pross has called *dynamic kinetic stability* (DKS) (Pross 2012); in other words, the perdurance of replicative molecules and autocatalytic cycles, which is an essential condition of self-organization of chemical systems. DKS is a particular type of stability, uncommon in non-biological chemistry, but peculiar to biochemical systems. In the words of Pross, life is a kinetic state of matter. These new concepts also require new experimental approaches. Classical prebiotic chemistry may not serve in this context and it is necessary to develop what some have called *systems chemistry*, the study of complex mixtures of molecules that interact with each other and that can give rise to emergent properties through mutual catalysis or the emergence of more complex structures. In contrast to classical prebiotic chemistry, systems chemistry shall eventually provide empirical instances of minimal systems exhibiting one or more characteristics usually associated to living cells.

Thus life is manifested through this peculiar chemistry with unique structural and organizational features. So we ask ourselves: what is universal (in terms of necessary) and what is contingent (in terms of being subject to randomness) in the structure and organization of life?

3.4 Extant Is a Clue for the Extinct

For Morowitz (1992), universality is synonymous with antiquity and he proposes that the characteristics shared by all life on Earth would be present in the common ancestor of all things alive today. That is, all life forms preceding us—during the almost 4-billion-year long evolutionary history of Earth—would have had the same fundamental biochemical characteristics. This exercise of meticulously tracking the descending branches of the tree of life to the common core may be obscured or hampered by phenomena such as horizontal gene transfer. This may affect the idiosyncratic metabolic details of some specific taxonomic groups. The bulk of the most fundamental shared features must be the result of genuinely Darwinian vertical inheritance, which would enable us to retrospectively reconstruct the metabolic and molecular aspects of our remote ancestors. Thus we would be able to establish a clear relationship between extant life and extinct life, although fossil remains of these early unicellular ancestors provide little or no information about their biochemical composition or even their geological dating remains controversial. Technically today it is possible to resurrect extinct proteins and study their function.

Apart from present and past life, which is genealogically related and subject to theoretical and experimental study, we can also consider potential life forms, exploring chemical alternatives to known processes, something that—as we have said—has already been done in some laboratories. In fact we have explored alternative genetic alphabets and tested the properties of genetic materials that would have theoretically been possible in parallel evolutionary histories but never materialized (i.e. life as we do not know it, see Chap. 4).

3.5 Individual and Collective Life

Several authors have emphasized the need to characterize the unique aspects of life. We can take the concept of autopoiesis proposed by Humberto Maturana and Francisco Varela as a guide. According to these authors a molecular autopoietic system represents the minimum living organized structure, understood as a network of the production processes (synthesis and destruction) of components, so that they continuously regenerate the network that produces them and constitute the system as a separate unit and clearly differentiated medium.

Despite its level of abstraction, we can easily relate the concept of autopoiesis to the fact that living things are open-ended thermodynamic systems, far from being balanced, with metabolic networks inside and a natural boundary separating the metabolic interior from the exterior (membrane), which is, in turn, a product of the metabolic network.

In principle, an autopoietic entity can grow and reproduce by division. These phenomena have been simulated in simple chemical experiments. According to this definition, a self-reproducing vesicle that can generate components of its membrane from the inside is alive.

But, clearly, autopoiesis does not capture all the fundamental aspects of life: paraphrasing Richard Dawkins one may state that all living things are part of a historical river where digitized information flows in the form of DNA polymers passing information from one generation to the next. We are the result of a long evolutionary history, in which time and chance have shaped our genomes.

This historical and collective aspect of life is captured by the definition used by NASA's exobiology programme since 1992, an adaptation of Carl Sagan's so-called "genetic definition": "Life is a self-sustaining chemical system capable of Darwinian evolution" (i.e., by natural selection). The fundamental properties of the system are, therefore, according to the neo-Darwinian paradigm of variability and heritage, in permanent contrast to environmental conditions. According to this definition, a population of self-replicating RNA molecules is alive.

To further specify the kind of organization showing these fundamental aspects of individual and collectively-historical living beings, the following definition was proposed: *living beings are autonomous systems with open-ended evolutionary capacities* (Ruiz-Mirazo et al. 2004).

An autonomous system is far from equilibrium and maintains itself by an organization based on a set of thermodynamic couplings between the internal processes of self-construction and interaction with its environment. Furthermore, the capacity of open-ended evolution is the potential of the system to reproduce its functional dynamics and cause a limitless variety of equivalent systems without a higher level of organizational complexity (apart from the restrictions imposed by finite resources and certain universal physicochemical laws).

This definition not only allows us to discuss what type of chemical systems and what minimum requirements are needed to show these fundamental properties but, following Oparin's tenets, enables us to trace a genealogical explanation of the definition itself (Peretó 2005). From the first approach, experimental protocols can be derived to test samples of putative life, while from the second approach hypothetical milestones can mark the transition from prebiotic to biological evolution, through proto- or infra-biological stages.

3.6 To Be Alive or to Be a Living Being—That Is the Question

We have endeavoured to bring together under one definition the ability to survive, to remain *alive* in certain conditions and counteract individual death by an incessant flow of matter and energy, and the concept of a *living being*, that is, belonging to a species with an evolutionary history, under the threat of collective extinction. Therefore, we are subject to two biographies: the individual, which unfolds in an ontogenetic programme, and the collective, a consequence of phylogenetic descent.

The theoretical biologist Tibor Gánti introduced the idea of "life criteria" and distinguished between absolute (or real) and potential criteria of a living being (Gánti 2003). These are not definitions of a live being but rather a taxonomic demarcation to classify objects. This enables us to separate the different states properly (*being* and *to be alive*) and to tackle common borderline cases: Are sterile animals or non-dividing cells alive? What about viruses?

Among the absolute criteria of a living being, Gánti speaks of individuality (a living being as a unit, which has properties that are not exhibited by its parts), metabolism, stability when faced with environmental changes, possession of an information subsystem and regulation of their internal processes. For Gánti, potential criteria of living things are: growth, reproduction, heredity and mortality. Szathmáry has disseminated Gánti's ideas and—refining his proposal—has suggested *units of life* (that meet the absolute criteria of Gánti) and *units of evolution* (that qualify to evolve by natural selection). At the intersection between units of life and units of evolution, we find any system that meets both absolute and potential criteria (e.g., free-living bacteria). But we can find cases of units of life that cannot evolve (a sterile animal) or evolutionary units that do not meet some absolute criteria (a virus).

In genealogical terms, we have proposed that, in their earliest origins, the first self-supporting chemical systems must learn to 'be' alive to enable them to incorporate evolution by natural selection and to 'become' living beings (Ruiz-Mirazo et al. 2004; Peretó 2005).

3.7 Awakening from the Cartesian Dream

For René Descartes, an animal is a machine, literally. No doubt Cartesian mechanisms, allied with the reductionism inherent to molecular biology, have given rise to remarkable scientific results. But we must recognize the limitations of these conceptions. Living beings, for the fundamental reasons discussed below, are not like machines and a full understanding of how they operate will not be gained by analysing their parts, not even by trying to recompose them crudely.

In his first work on the origin of life, Oparin proposed a list of properties manifest by living beings (defined organization, metabolism, reproduction and excitability), none of which is unique to them alone, as they can be found in inanimate nature.

Oparin concluded that "the specific peculiarity of living organisms is only that in them there have been collected and integrated an extremely complicated combination of a large number of properties and characteristics which are present in isolation in various dead, inorganic bodies". For the Russian biochemist, to find the conditions on early Earth that facilitated the special and specific combination of properties we see in living things would be to explain the origin of life (Lazcano 2008).

Another point of view is to consider that the nature of life is distributed in different subsystems, none of which exhibits all the properties attributed to living beings. These properties emerge only when all subsystems are combined and coupled. According to Gánti's chemoton model and Szathmáry's terminology (Gánti 2003) we can establish three infra-biological (or supra-chemical) subsystems: *metabolism, boundary* and *template*. As we will see later (Chap. 4), experimental simulation allows for certain combinatorial possibilities among them, enabling us to explore possible early stages of cellular evolution and even to speculate on the grade of "livingness" of each step in that protobiological era (for further discussions see Peretó 2005).

We cannot end this chapter on the essence of life without referring to Robert Rosen's ideas. Although his theory of (*M,R*) systems (metabolism, repair or, rather, replacement) is highly abstract and difficult to grasp, we should thank Athel Cornish-Bowden and co-authors for their endeavour to explain his interesting ideas (Letelier et al. 2011). The idea of autopoiesis emphasizes an organism's structural organization and the existence of a physical barrier separating the inside from the outside, while the idea of (*M,R*) systems highlights the logic of organization expressed in mathematical terms.

One of the key observations is obvious, although it is rarely taken into consideration: all the enzymes required for an organism to live must be produced by the organism itself (from this perspective the inert character of viruses is again in contrast). This leads to the so-called metabolic circularity or, in more modern terms: the metabolome is a product of the proteome and this, in turn, is a product of the metabolome (sic).

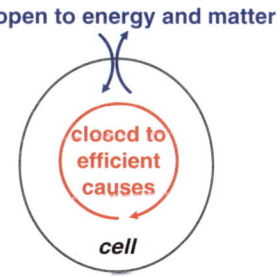

Fig. 3.1 Living systems are closed to efficient causes but open to energy and matter flows (they are open thermodynamic systems). The formulation of Robert Rosen regarding recursivity or metabolic circularity in cells is one of the clearest distinctions between living systems and artificial machines

In fact, enzymes have a finite life and must be constantly re-synthesized from components of the organism itself (from within). This immediately shows us the inappropriateness of comparing organisms and machines. All man-made artefacts need outside agents to design, manufacture and maintain them. In an organism, the replacement of components (and construction to a great extent) is an internal function, which does not require outside help. Rosen put it as follows: "A material system is an organism if and only if it is closed to efficient causation" (Fig. 3.1). There is, therefore, an essential and fundamental difference between organisms and machines that cannot be ignored in the context of synthetic biology (Morange 2012a).

References

Bedau MA, Cleland CE (eds) (2010) The nature of life. Classical and contemporary perspectives from philosophy and science. Cambridge University Press, Cambridge

Gánti T (2003) The principles of life. With comments by J Griesemer and E Szathmáry. Oxford University Press, Oxford

Lazcano A (2008) What is life? A brief historical overview. Chem Biodivers 5:1–15

Letelier JC, Cárdenas ML, Cornish-Bowden A (2011) From L'Homme Machine to metabolic closure: steps towards understanding life. J Theor Biol 28:100–113

Morange M (2003) La vie expliquée? 50 ans après la double hélice. Odile Jacob, Paris. English translation: Life explained, 2008. Yale University Press, New Haven

Morange M (2012a) Synthetic biology: a challenge to mechanical explanations in biology? Perspect Biol Med 55:543–553

Morange M (2012b) The recent evolution of the question "What is life"? Hist Philos Life Sci 34:425–436

Morowitz HJ (1992) Beginnings of cellular life. Metabolism recapitulates biogenesis. Yale University Press, New Haven

Peretó J (2005) Controversies on the origin of life. Int Microbiol 8:23–31

Peretó J, Bada JL, Lazcano A (2009) Charles Darwin and the origin of life. Orig Life Evol Biosph 39:395–406

Pross A (2012) What is life? How chemistry becomes biology. Oxford University Press, Oxford

Ruiz-Mirazo K, Peretó J, Moreno A (2004) A universal definition of life: autonomy and open-ended evolution. Orig Life Evol Biosph 34:323–346 (Reprinted in Bedau and Cleland (2010))

Ruiz-Mirazo K, Peretó J, Moreno A (2010) Defining life or bringing biology to life. Orig Life Evol Biosph 40:203–213

Tawfik DS (2010) Messy biology and the origins of evolutionary innovations. Nat Chem Biol 6:692–696

Chapter 4
Strategies for Making Life

Abstract Synthetic biology is a multifaceted discipline and the pathways towards an artificial cell are diverse. Top-down strategies seek simplification of genomes, their chemical synthesis and transplantation into a cell chassis. In the long term, scientists hope to have genomic platforms to reinvent metabolic networks capable of producing molecules of biotechnological interest. On the other hand, a bottom-up strategy relies on the chemical implementation of fundamental concepts such as self-reproduction, self-replication and self-maintaining systems. In addition to the artificial synthesis of simplified genomes and protocells, some scientists explore xenobiology, or making life as we do not know it, for example, with different genetic alphabets or with artificially designed metabolic pathways.

> *After days and nights of incredible labour and fatigue,*
> *I succeeded in discovering the cause of generation and life;*
> *nay, more, I became myself capable of bestowing animation*
> *upon lifeless matter.*
> Mary Shelley *Frankenstein, or the modern Prometheus* (1818)

> *The eobiont I intended to create had to be the simplest possible*
> *form of independent life [...] I have broken down a*
> *metaphysical border by erasing the boundary between*
> *chemistry and biology.*
> Harry Mulisch *The Procedure* (1998)

4.1 Frankenstein and Werker: Two Strategies to Make a Living Being

Advanced scientific knowledge has always inspired literature, in fact, more often than it might seem at first glance. In particular, the speculative exercise of potentially making a living being, a human being, is a legendary theme. The golem of premodern Jewish tradition or the homunculus of Paracelsus, are both expressions of the eternal tension between nature and human creations, also reflected in works of art. The golem arises from mud and magic spells, to give rise to a subhuman

M. Porcar and J. Peretó, *Synthetic Biology*,
SpringerBriefs in Biochemistry and Molecular Biology,
DOI 10.1007/978-94-017-9382-7_4

being deprived of our most genuine capacity: speech. We must destroy it daily, before it develops into an uncontrollable and dangerous force. However, the homunculus, the alchemists' artificial man, is a being comparable to natural human perfection, and represents the pinnacle of our technological and creative power. The very idea of the almost unlimited powers unleashed by revealing nature's innermost secrets, and the negative consequences this might bring, gave rise to a plethora of classical mythological figures, like Adam and Eve, Prometheus or Faust. In all cases, humankind's craving for knowledge and natural thirst for discovery are portrayed linked to fear of the unknown and to the inevitable disaster that comes from usurping that which is reserved for the gods alone.

When we talk of creating life, a certain literary character immediately comes to mind: Dr. Victor Frankenstein, the protagonist of the novel by Mary Shelley. This work quickly gained favour with the public and is well known thanks to the more or less faithful theatrical and film versions. Hence, it has been the subject of numerous studies by literary critics, philosophers and historians. However, few know that in this story, published in 1818, Shelley based the figure of Frankenstein, and his quest to make a human, on the most advanced research of her time. Frankenstein is a scientist with a strong background, who is convinced that there is nothing magical or supernatural about life. Despite having read the great alchemy treaties during his youth, his university teachers later convinced him that physics and chemistry provide more consistent and plausible explanations for the functioning of living beings. Thus, young Frankenstein's chemistry professor, Mister Waldman, comments that "the ancient teachers of this science [the alchemists] promised impossibilities, and performed nothing. The modern masters promise very little [but] have indeed performed miracles". Miracles, figuratively speaking, because—according to Waldman—these scientists "penetrate into the recesses of nature, and show how she works in her hiding places". From the day he met Waldman, Frankenstein makes his sole occupation "natural philosophy, and particularly chemistry, in the most comprehensive sense of the term".

When Mary Shelley fist published her famous novel *Frankenstein, or the Modern Prometheus* in 1818, she made it clear that what she recounted in the book was not considered impossible by scientists of her time. The author based the story on scientific progress made by the likes of William Harvey, who discovered blood circulation; Erasmus Darwin (Charles Darwin's grandfather), an early evolutionist who was quoted in the book's foreword; Sir Humphry Davy, an illustrious English chemist; or Luigi Galvani, an Italian physicist and doctor who studied the effects of electricity on animal health. After much effort and sacrifice, Frankenstein will "in discovering the cause of generation and life" recognize its potential for "bestowing animation upon lifeless matter". So, Dr. Frankenstein will be able to revitalize a lifeless body with fragments of dead matter that was once living. The consequences that this brought upon him are well known.

One hundred and 80 years later in *The Procedure*, the Dutch novelist Harry Mulisch narrated the story of Victor Werker, a scientist capable of making very simple cells ('eobionts') from chemical compounds, such as certain clays. Mulisch based his story on actual research on the origin of life. Obviously, both novels show us the fate met by those 'playing God'. But what matters here is that both Frankenstein and Werker are serious and rigorous scientists pursuing the same goal, but based on different paradigms.

So, paraphrasing Evelyn Fox Keller, Frankenstein pursues the artificial synthesis of life—an artificial procedure to restore life to a being that once possessed it—while Werker wishes to synthesize artificial life—an artificial life of simple characteristics resembling that which emerged on the primitive Earth. In other words, one can endeavour to create living beings by using parts of other existing beings (*top-down*) or synthesize basic forms of life by chemically implementing key concepts, such as autopoiesis or self-replication (*bottom-up*).

4.2 À la Frankenstein: Artificial Synthesis of Life or the Top-Down Approach

We could define this branch of synthetic biology as the genetic and metabolic engineering of the post-genomic era, in other words, biological engineering put into the context of systems biology. Thus, a fundamental difference with the classical genetic engineering approach is that we now know the full extent of the genetic landscape of a cell, and can thus infer metabolic networks, genetic circuits or protein interactions. A previously unimplemented stage is the one involving computational simulation prior to experimental intervention. Indeed, quantitative theoretical computer models and simulations play a significant role in this type of approach to synthetic biology.

One of the most striking examples in this area is the proposal by MIT engineers who launched the idea of standardization and cataloguing of so-called interchangeable components known as *Biobrick* parts (see Chap. 6). These can be combined to design new genetic and metabolic circuits, new properties in host cells, which act as a mere 'chassis'.

A basic criticism to be made of this approach is the misplaced comparison between cells and computers. As we saw in Chap. 3, there are profound reasons marking the difference between the functioning of machines and of living cells. Furthermore, it is not to be expected that biological 'parts' (like proteins such as transcription factors, or promoter sequences and other elements) have a totally reliable and predictable behaviour in any context. That is, it is more than doubtful they are indeed completely interchangeable. A protein may act in a certain way within the context in which it has been characterized, but manifest other properties or functions in a different context. An important fact emerging from postgenomic analysis of cellular life, it is that many proteins have more than one recognized function (*moonlighting* or multifaceted). In most cases we rarely ascribe a sole function. But what it is more intriguing, as discussed in Chap. 3, proteins can also show *promiscuity*, secondary functions not subject to selection and that may become important under changing environmental conditions. These properties are

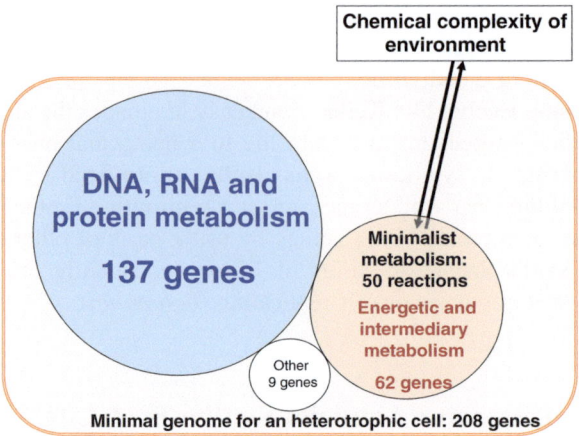

Fig. 4.1 A proposal of minimal genome and minimalist metabolism (based on Gil et al. 2004; Gabaldón et al. (2007)). Albeit we could expect that the minimal genetic machinery to be almost universal, the complexity of a minimalist metabolism is ecologically-dependent and will be an inverse function of the chemical complexity of the environment

directly derived from proteins' intrinsic flexibility and have undoubtedly played an important role in evolution (Tawfik 2010). Therefore, one cannot rule out the emergence of new and unexpected properties when gathering 'parts' that were characterized in different contexts.

Another way to emulate Frankenstein at the cellular level is the path taken by John Craig Venter and colleagues (see Chap. 5 for further details). They wish to define the minimal genome to which one may add packets of information, which will reconfigure cell functions with a biotechnological aim. Without doubt, the most immediate task is to list genes that constitute the minimum set of genetic information necessary and sufficient for cell life under certain conditions.

Parasitic bacteria or intracellular symbionts have undergone a remarkable genomic reduction process. Comparative genomics has revealed minimum gene numbers of around 200, corresponding to a bacterium in a chemically rich environment. This genome contains sufficient information for a minimum stoichiometrically consistent heterotrophic metabolism (Fig. 4.1).

The complexity of a metabolic network depends on the complexity of the environment, meaning that a *single* minimal genome does not really exist. Rather, there are as many as there are environments imaginable (Fig. 4.1). The idea of a minimal cell, whose simplicity is a function of ecological complexity, was introduced by Morowitz. Moreover, the study of more endosymbiotic bacteria (and bacterial consortia) can enrich our knowledge of how evolution has reached a minimal genome in very different environments. It is also of special interest to understand how free-living cells have reached minimalist solutions regarding how to deal with external sources of matter and energy. This information can be very useful in the context of a genomic oriented synthetic biology.

Venter's group has identified the essential genes in one of the smallest genomes that exist in culturable bacteria, *Mycoplasma genitalium*. A patent was thus filed for these 381 genes, a fact that caused a great stir. Furthermore, the same group was able to chemically synthesize the bacterium's entire genome of 582,970 base pairs, a feat more technological in nature than scientific (see Chap. 5). This is because if one wishes to produce a genome de novo, techniques for chemical synthesis of polydeoxynucleotides (i.e. DNA) must be developed, enabling production at an even greater speed, in the most reliable and inexpensive way possible.

What else is there to do? Not only do we have to construct a genome, assuming we know the list of genes, but we must also know which order to put them in, what regulatory sequences exist and their positioning, etc. Additionally, the genome must be inserted inside a cell lacking its own genome, which will act as platform for its expression. Venter's group has managed a 'genome transplant' in a very special case: the genome of *M. mycoides* (1.08 million bp) was inserted inside cells of *M. capricolum*. The result was a synthetic species, *M. mycoides* JCVI-syn1.0. It is unlikely that this procedure can be generalized to other cell systems. Among other technical details, one must remember that Mycoplasmataceae are the only bacteria lacking a cell wall, thus facilitating the entry of foreign DNA. In Fig. 4.2 we present a generalized scheme for the whole process of inserting a synthetic genome in a cell chassis.

Which is the ultimate goal? If in the end we have a cellular system supporting and expressing chemically synthesized genomes, we could design combinations of genes that confer properties to the artificial cell of biotechnological or biomedical interest. Obviously, to reach this goal, much theoretical and experimental work lies ahead. In any case, we should try to overcome this somewhat overly simplistic view

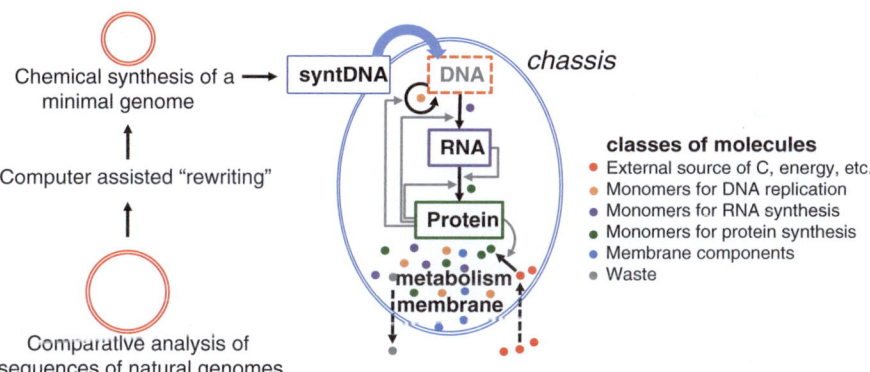

Fig. 4.2 Genome transplantation. A minimal synthetic genome, based on biological information, computing "rewriting" and chemical synthesis is introduced in a genome-free cell (chassis) with the minimal chemical devices and components to kick-start metabolic networks. In its simplest form, a chemical machinery is necessary to transform a external source of matter and energy into the diversity of cell precursors, namely, monomers for the synthesis of nucleic acids (ribo- and deoxyribo-nucleotides), monomers for the synthesis of proteins (amino acids), membrane components, as well as other essential chemicals (e.g., vitamins or enzyme cofactors)

of biological complexity. Indeed, we should remember that we *cannot put every-thing down to genes*. Cellular complexity is governed by an unpredictable deter-minism: the cell itself and its interaction with the environment still hold many enigmas. And it seems that even with the maximum of information on a small biological system, we are still far from understanding its functioning (see discussion on the detailed molecular characterization of *M. pneumoniae* in Chap. 5).

4.3 À la Werker: Synthesis of Artificial Life or the Bottom-up Approach

We saw in Chap. 2 that pioneers of the scientific study into the origin of life, such as Oparin, established the ultimate goal of this research as the ability to make cells in the laboratory. John B. S. Haldane, a British scientist who also provided us with very insightful reflections on the origin of life, stated "it may be that artificial life of a simple character will be made in the laboratory long before we understand the process going on inside the cells" because "we know enough to say that the process is not mysterious but merely complicated". Thus, a long scientific tradition has followed this strategy, both theoretically and experimentally: from Oparin's and Haldane's postulates to the prebiotic chemistry founded by Miller, up to models like Gánti's *chemoton* (i.e., a theoretical model of functional coupling of the three suprachemical/infrabiological subsystems, see Chap. 3). Thus, this model can provide guidance for the construction of artificial living systems or something very similar. The opinion has long been held that we can advance in the synthesis of life from simpler materials, as a way of exploring chemistry's ability to organize itself as biological systems, something that happened spontaneously on Earth in the distant past. This *bottom-up* approach is based on the conviction of being able to implement fundamental principals in the test tube, such as the autocatalytic pro-cesses of molecular replication, reproduction of lipid vesicles or self-perpetuating chemical networks.

Let us accept the three basic infrabiological subsystems mentioned above (see also Chap. 3)—metabolism, boundary and template. The big challenge is to chemically configure them and establish combinations between them. In order of increasing success to date the combinations could be: chemical networks that sustain self-replicating polymers (so far lacking examples), vesicles derived from simple reactions located within them, or vesicles containing self-replicating poly-mers. Experimental systems published to date are very close to the third type. We could even establish a hierarchy: protocells that are self-maintained, self-replicating or with the ability to evolve, depending on the degree of chemical complexity implemented. Undoubtedly, the most difficult challenge is to find how the three subsystems, which one can simulate separately in the laboratory, engage with each other harmoniously. To answer this key question on the origin of life would be to hit the bull's eye.

At any rate, if we accept the Oparinian tradition that the emergence of life on Earth followed an evolutionary, continuous transition it is meaningless to draw a sharp border between nonliving and the living. With Antonio Lazcano we think that it is a waste of time to discuss when exactly did life start. Instead we should engage in understanding how certain properties found in inanimate (prebiotic, supra-chemical, infrabiological) systems are articulated in primitive living systems (Lazcano 2008).

In this vein, Jacques Reisse and coworkers have proposed the application of fuzzy logic to prebiotic chemistry and a scale of "livingness" or *life index* to characterize the emergence of complex systems in early Earth on their way from inert matter to the primitive cells (Bruylants et al. 2010). They propose to qualify as life index 0 chemicals that are found in both the inorganic world (e.g. meteorites, comets) and the living cells, for instance water or alanine. Then the life index will reach 1 in cells with a complexity degree similar to that of modern prokaryotes. Anything in between (proteins, genomes, ribosomes, etc.) will show a life index less than 1 but higher than 0. Our infra-biological (or supra-chemical) subsystems and their combinations (Fig. 4.3) also deserve a life index between 0 and 1, the closest the chemical system to the simplest cell, the higher the life index. These authors propose this scale to redefine the aims and discipline borders of prebiotic chemistry, a field that some authors prefer to rename and update as systems

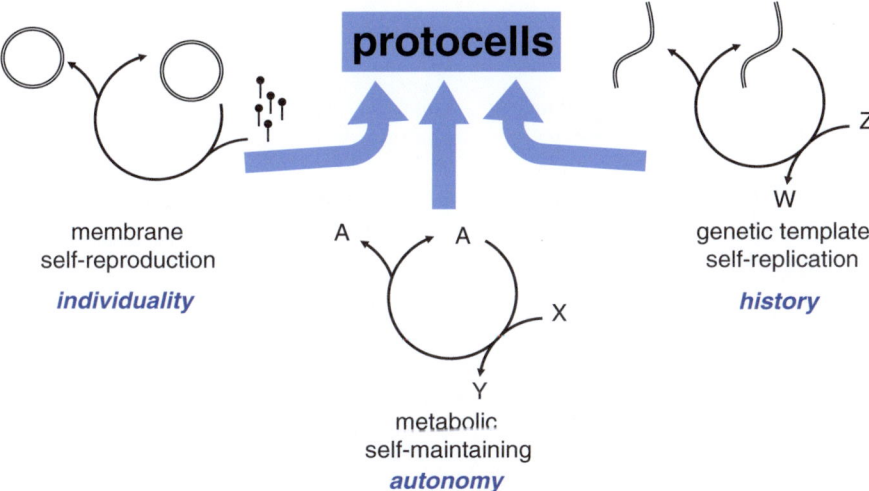

Fig. 4.3 Three basic infrabiological (or suprachemical) autocatalytic subsystems relevant for the origin and artificial synthesis of life (adapted from Peretó 2012). Self-reproduction of lipid vesicles, self-maintaining of metabolic networks, and self-replication of genetic polymers constitute the three basic sub-systems of a minimal biological system. The coordinated assembly of the three subsystems is the ultimate goal of a bottom-up approach of synthetic biology, also relevant in the context of the empirical exploration of the origin of life (i.e. prebiotic and systems chemistry)

chemistry and that could be epistemologically indistinguishable of the bottom-up approach to synthetic life.

Jack W. Szostak, David P. Bartel and Pier Luigi Luisi are three important players in the chemical implementation of two of the three infrabiological subsystems (specifically, ribozymes as model replicative polymers and vesicles as model simple cells). In 2001, at the request of the journal *Nature*, they published a real research programme for the synthesis of artificial life, bottom-up (Szostak et al. 2001). The experimental foundations of this achievement are already in place, but we need to know much more about both RNA evolution within the test tube, and about the biophysics of membranes and vesicles.

In essence, it would take the production of self-reproducing vesicles containing at least two ribozymes: one with a RNA-dependent RNA-polymerase activity (i.e., capable of copying the RNA template, both to itself and to a second ribozyme), and another with an activity enabling it to catalyse the synthesis of the molecules constituting a vesicle membrane. This scenario could represent one of the simplest synthetic cell machinery inspired by the notion of a primitive RNA world (Szostak et al. 2001). In Fig. 4.4 we compare this possible artificial cell with other minimal cells.

Just to highlight one remarkable achievement on protocell development, in 2011 a Japanese team led by Tadashi Sugawara reported an experiment in which, for the first time, a link between vesicle reproduction and informational replication was established (Kurihara et al. 2011). This work combines two previous technical achievements: self-reproduction of the compartment and in vitro replication of an

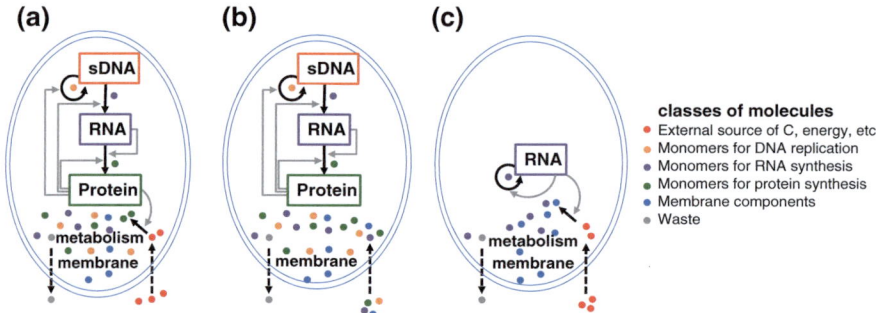

Fig. 4.4 A diversity of minimal artificial cells. **a** A synthetic cell with a transplanted synthetic genome in which a minimalist metabolism has been encoded allowing the transformation of external sources of matter and energy in the minimal cell components: monomers for the synthesis of nucleic acids (ribo- and deoxyribo-nucleotides), monomers for the synthesis of proteins (amino acids), membrane components, as well as other essential chemicals (e.g., vitamins or enzyme cofactors) (cfr. Fig. 4.2). **b** Some authors imagine the possibility of a permeable membrane and the supply of all cell components from the environment leading to a hypothetical minimal cell without metabolism. **c** A further simplification is represented by a ribocell where all the genetic information is encoded in a self-replicating RNA containing the minimal instructions for the metabolic conversion of external molecules in at least two cell metabolites, i.e., monomers for RNA synthesis and membrane components (see Szostak et al. 2001)

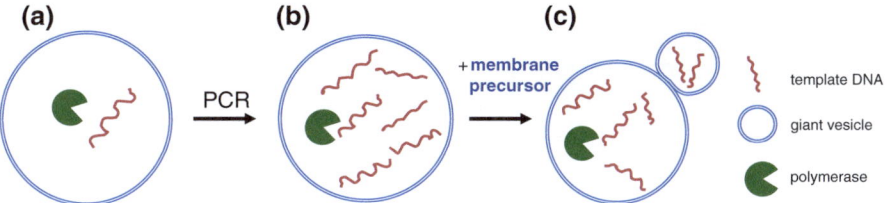

Fig. 4.5 Chemical link between amplification of DNA and self-reproduction of vesicles. **a** Encapsulated template DNA and PCR reagents in giant vesicles. **b** PCR of template DNA takes place inside giant vesicles. **c** After adding molecular precursors of membrane components, growth and spontaneous division of vesicles, as well as distribution of DNA in the daughter vesicles is observed. Amplification of DNA accelerates vesicle self-reproduction. Based on Kurihara et al. (2011)

informational molecule. In their work, the team observed the growth and sponta- neous division of giant artificial vesicles containing DNA, which was not only transferred to the daughter giant vesicles, but proved able to accelerate the division of the giant vesicles following its amplification by the Polymerase Chain Reaction (PCR) (Fig. 4.5). This link between polymer replication and vesicle reproduction sheds light on the origin of cells and paves the way towards a bottom-up fully functional protocell.

4.4 Synthesizing Life as We Do not Know It

A non-conformist approach to metabolic engineering has recently emerged. Instead of introducing changes in the expression, activity and/or organisation of individual enzymes in natural pathways, including the introduction of heterologous activities to produce either inherent or non-inherent products in a given metabolic network, imaginative scientists change the entire stoichiometric design of a pathway modi- fying fundamental aspects of metabolism. For instance, the universal mode of hexose utilization by cells up to acetyl-CoA shows the mandatory stoichiometric loss of 33 % of carbon as CO_2 (Fig. 4.6a). Thus, six carbons in hexose are con- verted in two (two-carbon) acetates (i.e. acetyl-CoA) and two CO_2. James C. Liao and coworkers at the University of California Los Angeles have designed a non- oxidative glycolysis (NOG) by means of combining, both in vitro and in vivo in *E. coli*, several heterologous activities with the aim of altering the universal stoi- chiometry of hexose degradation to achieve 100 % of carbon utilization (Fig. 4.6b). Hence, the artificially designed NOG allows the complete carbon conservation in sugar metabolism with the conversion of one hexose in three acetyl-CoA molecules (Bogorad et al. 2013). Other pathways have also been object of reconfiguration: a reverse glyoxylate cycle allows the conversion of C4 molecules in C2 molecules

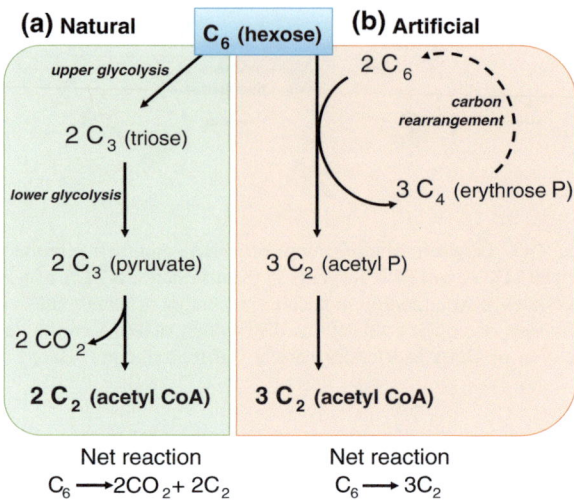

Fig. 4.6 Stoichiometric scheme of natural glycolysis (**a**) compared with artificial non-oxidative glycolysis, NOG (**b**) (based on Bogorad et al. 2013). Natural glycolysis shows a universal stoichiometric loss of two molecules of CO_2 per one hexose molecule consumed, whereas in the artificially designed NOG all six carbons of hexose are transformed in acetyl CoA

without carbon loss, whereas a reverse β-oxidation of fatty acids opens the door for more efficient pathways of synthesis of fuels and chemicals of industrial interest.

Furthermore, researchers have explored the potential of changing the spatial location of enzymes. Natural selection has favoured metabolic channelling (i.e., the physical association of consecutive enzymes in a pathway) when chemical intermediates are labile, show high and spontaneous reactivity or easily escape the cell by diffusion. Metabolic engineers have mimicked evolution artificially inducing the association of enzymes using scaffolds (of peptide or nucleic acid nature) with specific binding sites for the accordingly modified enzymes –originally, a project presented by an iGEM team (see Chap. 6). Synthetic scaffolding, with different architectures, has shown successful increases in the production of some metabolites. In the near future synthetic subcellular compartments will allow the colocalization of metabolic enzymes under optimal environmental conditions.

Following Benner et al. (2011) discussion on the diversity of meanings for "synthetic biology", still there is a stronger way to depart from the engineering approach to biology. Thus, engineers seek the use natural molecular parts from living systems to design new devices performing things that are not done by natural systems. But the approach of chemists like Benner is the opposite. They try to use unnatural molecular parts to perform things that are done by natural biological systems. This is also known as xenobiology, an emerging research topic aiming at the construction (for instance) of functional alternative nucleic acids. After initial failures and by using non-deoxyribose sugars, alternative genetic polymers based on simple nucleic acid architectures not found in nature, xeno-nucleic acids (XNAs)

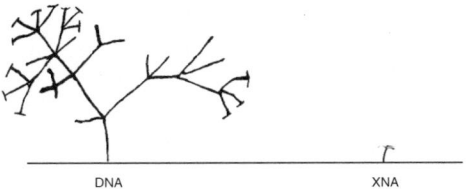

DNA   XNA

Fig. 4.7 A new life tree may be just sprouting close to our old Darwinian life tree (based on an idea of Markus Schmidt)

capable of evolution have been reported. Synthetic life with more than four letters in the genetic alphabet (see, e.g., Malyshev et al. 2014) or more than twenty amino acids in the protein composition is possible in engineered bacteria (discussed by Benner 1994). Now the big question is not whether a self-maintaining artificial chemical system capable of evolution will be possible, but rather when would that happen.

Interestingly, this "chemically parallel life" (Fig. 4.7) is an emerging research field that was first proposed as the ultimate biosafety tool, mainly because of its supposed incompatibility with standard DNA- and RNA-based life (Schmidt 2010). However, concerns have now arisen on the possibility of interactions of xeno-organisms with today's living forms.

But, let's insist once more, such artificial biology will expand (it is really expanding it now) our understanding of life (Ruiz-Mirazo and Moreno 2013). We remind the sentence written by Loeb more than 100 years ago, "we must either succeed in producing living matter artificially, or we must find the reasons why this is impossible" (Chap. 2). This remarkable epistemological position echoes the one proposed by Benner when he defends the value of failure and the importance of analysing the reasons of failure to learn more on the behaviour of chemical systems.

At any rate, engineers, chemists, synthetic biologists altogether are united by the same intellectual Oparinian thread: "the artificial synthesis of living things [is] not [an] unattainable goal".

References

Benner SA (1994) Expanding the genetic lexicon: Incorporating non-standard amino acids into proteins by ribosome-based synthesis. Trends Biotech 12:158–163

Benner SA, Yang Z, Chen F (2011) Synthetic biology, tinkering biology, and artificial biology. What are we learning? C R Chimie 14:372–387

Bogorad IW, Lin TS, Liao JC (2013) Synthetic non-oxidative glycolysis enables complete carbon conservation. Nature 502:693–697

Bruylants G, Bartik K, Reisse J (2010) Is it useful to have a clear-cut definition of life? On the use of fuzzy logic in prebiotic chemistry. Orig Life Evol Biosph 40:137–143

Gabaldón T, Peretó J, Montero F, Gil R, Latorre A, Moya A (2007) Structural analyses of a hypothetical minimal metabolism. Philos Trans R Soc Lond B Biol Sci 362:1751–1762

Gil R, Silva FJ, Peretó J, Moya A (2004) Determination of the core of a minimal bacterial gene set. Microbiol Mol Biol Rev 68:518–537

Kurihara K, Tamura M, Shohda K, Toyota T, Suzuki K, Sugawara T (2011) Self-reproduction of supramolecular giant vesicles combined with the amplification of encapsulated DNA. Nature Chem 3:775–781

Lazcano A (2008) What is life? A brief historical overview. Chem Biodivers 5:1–15

Malyshev DA et al (2014) A semi-synthetic organism with an expanded genetic alphabet. Nature 509:385–388

Peretó J (2012) Out of fuzzy chemistry: from prebiotic chemistry to metabolic networks. Chem Soc Rev 41:5394–53403

Ruiz-Mirazo K, Moreno A (2013) Synthetic Biology: challenging life in order to grasp, use, or extend it. Biol Theor 8:376–382

Schmidt M (2010) Xenobiology: a new form of life as the ultimate biosafety tool. BioEssays 32:322–331

Szostak JW, Bartel DP, Luisi PL (2001) Synthesizing life. Nature 409:387–390

Tawfik DS (2010) Messy biology and the origins of evolutionary innovations. Nat Chem Biol 6:692–696

Chapter 5
Synthetic Biology in Action

Abstract Last decade has witnessed remarkable advances towards the engineering of life. The examples range from the design of an efficient cellular factory for the semi-synthesis of the antimalarial drug artemisinin, to the chemical synthesis of chromosomes, both bacterial and eukaryotic. In parallel, advances in the deep characterization of cell machineries in the simplest cells show that we are very far of fully understanding the regulation of metabolic and genetic circuits. Biological emergent properties and noise may suppose an obstacle for predictive design. Besides the obvious biotechnological benefits of synthetic biology, the path towards the artificial cell will report new insights on the essence of life.

In this chapter, we will focus on the main milestones of research in synthetic biology. We will briefly discuss one of the foundational articles by Drew Endy, and then describe the evolution of this discipline through the main discoveries reported since 2005. From virus engineering to biofuel production and from artificial protocells to synthetic consortia, this overview will help to understand the potential and problems of engineering life. We will also highlight the trends in this discipline, which may take some important turns in the near future.

5.1 Virus and Malaria to Begin with

Although synthetic biology started a few years earlier, 2005 marks the start with several flagship reports on synthetic biology. Noteworthy was the review by Drew Endy on basic concepts of life engineering (Endy 2005), and the research report, also by Endy's team, on bacteriophage T7 rational modifications (Fig. 5.1). In his review, published in *Nature*, entitled *Foundations for engineering Biology*, Endy gathers the views of the principal synthetic biologists at that time (acknowledged in a long list at the end of the article). He stresses a very simple idea: bioengineering could be done otherwise; it could be done more easily by applying the same principles that have given way to the outstanding revolution of other technologies, which have subjected the natural world to human will, such as electronics. These three basic principles are: (i) the standardization or use of DNA in the form of

© The Author(s) 2014
M. Porcar and J. Peretó, *Synthetic Biology*,
SpringerBriefs in Biochemistry and Molecular Biology,
DOI 10.1007/978-94-017-9382-7_5

biological parts of predictable behaviour, not unlike the famous development of nuts and bolts; (ii) decoupling or reduction of the complexity of a system by splitting it both conceptually and industrially into separate tasks; (iii) abstraction, or the reduction of complexity by overcoming ad hoc characterization of biological parts—i.e., promoter strength—and using biological parts or circuits as blackboxes. It is important to highlight here the huge influence of this review article on the conception of life engineering, since this conception of the strictly applying engineering principles to synthetic biology has somehow guided the evolution of the topic since then.

Interestingly, two months before Endy's review was published, another outstanding report appeared in the journal *Molecular and Systems Biology*. Refactoring phage T7 is a visionary work in which a very simple system—a bacterial-*eating* small virus—was engineered with the ambitious goal of making it behave in a most rational—i.e., efficient—way (Chan et al. 2005). The researchers replaced around one third of the 39,937 bp genome with engineered DNA (which lacked non-essential overlapping genetic elements), and demonstrated the viability of the resulting virus that the authors named "T7.1". These results indicate that major design-based modifications of small genomes can be performed without loss of function. However, as the authors stated, the modified virus was not as fit as the original one: "T7.1 preserves polymerase-mediated genome entry and remains *relatively* independent of host cell physiology" (italics in the original report); "after 24 h at 37 °C, plaque sizes relative to the wild type were smaller for each of the chimeric phages".

One year after this important effort towards genome writing, another milestone in synthetic biology was accomplished by Jay Keasling's team at University of California Berkeley. These researchers engineered a whole metabolic pathway, in the budding yeast *Saccharomyces cerevisiae*, in order to produce artemisinic acid, which is an immediate precursor that can be easily chemically converted into the antimalarial drug artemisinin (Ro et al. 2006).

Until that major achievement, artemisinin, a sesquiterpene lactone endoperoxide, could only be obtained from the plant *Artemisia annua*, at rather low efficiencies. The development of a sophisticated synthetic metabolism in microorganisms boosted production rates and facilitated extraction, since artemisinic acid was accumulated outside yeast cells. After this initial success, several improvements in the process were reported. The most recent one by UC Berkeley scientists in cooperation with Amyris Inc., published in *Nature* (April 2013), describes the inclusion of a plant dehydrogenase and a second cytochrome that yielded a 15-fold improvement in artemisinic acid production, reaching 25 g/L. This high efficiency, along with improvements in the photochemical conversion of artemisinic acid to artemisinin and the open access of this technology will certainly revolutionize treatment of one of the most devastating diseases affecting tropical countries. Using this development and under a principle of "no-profit, no loss", the pharmaceutical company Sanofis has announced the production of up to 60 tons of artemisinin—some 150 million treatments—per year in 2014 at a price quite similar to the drug obtained from *A. annua*.

As in the first example of genome refactoring of phage T7, the original 2006 report on artemisinic acid biosynthesis is one of the most commonly cited works to demonstrate that synthetic biology does work. But even if the project was an overwhelming success, it should be noted that individual enzymatic modifications were not made following an orthodox engineering (i.e., standard-based) approach; indeed, a great deal of tinkering was required to make the whole system work as expected.

5.2 Playing God with a Chromosome?

The impact of these reports on the scientific readership was considerable, but it had a very limited reach on a broader audience. By contrast, experiments by John Craig Venter's team on cells with chemically synthesized chromosomes made headlines all over the world. The work was in fact the result of a long project, the first achievement of which was published in 2008 when the team reported the first chemically synthesized *Mycoplasma genitalium* chromosome (Young et al. 2008). This major technical achievement required the chemical synthesis of the 582,970-base pair genome of the bacterium in a synthetic copy with all but one of the genes of the wild-type strain: MG408 was disrupted for both blocking pathogenicity and allowing for selection. The construction was made from relatively small (5–7 kb) fragments, joined by in vitro recombination to yield medium-sized fragments (24, 72 and 144 kb), which were cloned in *E. coli*. Transformation-associated recombination to yield the complete genome was finally achieved in yeast. This first step towards a synthetic cell was complemented with another report by the same team, which published the technical procedure to transfer a genome from yeast to a receptive cytoplasm one year later (Lartigue et al. 2009). The authors transplanted the *M. mycoides* genome into another species host cell, *M. capricolum,* giving rise to a viable *M. mycoides* cell. The way was paved to the next and definitive work, published in *Science* in 2010 entitled "Creation of a bacterial cell controlled by a chemically synthesized genome" (Gibson et al. 2010). This work combines the preceding technical achievements to place a milestone not only in synthetic biology, but also in science in general: the production of the first living cell, "whose parent is a computer" (Venter in a press release 2010), *Mycoplasma laboratorium,* also known as Synthia. Both the authors' choice of the word "creation" and the ensuing media hype worldwide were strongly criticized by the scientific community. Nonetheless, this research work opened up a brand new scenario, in which synthetic genomes were synthesized, transplanted into genome-free host shells, and could subsequently take over the whole structure. This does not mean, though, that our understanding of the complexity of metabolic interactions and our ability to modify wild-type genes and networks enable us to truly "write genomes".

As proposed in Chap. 4, Venter's work corresponds to the so-called top down approach to artificial life. By using naturally streamlined genomes or by streamlining wild-type cells, this approach attempts to construct semi-synthetic, more or less engineered cells from a natural template cell. But as we discussed in Chap. 4, an even more ambitious yet long-term approach is the bottom up strategy, which focuses on simple constituents of a cell to build a truly synthetic cell from scratch. This approach uses protocells, artificial vesicles that somehow replicate natural cells' behaviour.

5.3 Cell Circuitry

Another imaginative approach to synthetic biocircuits is the implementation of connections between cells instead of between the molecular components inside the cells. The strategy consists of using cell consortia as basic modules to perform simple computations in a shared manner. Those systems could circumvent some of the problems encountered in the implementation of molecular circuits, such as unpredictability or randomness (Macía et al. 2012).

In this context, we should consider a report by Prindle et al. (2014) on post-translational coupling of genetic circuits, in which the authors set in place a protease competition system to engineer fast and tunable coupling of genetic circuits across both spatial and temporal scales. In other words, they developed a new approach to tune communication between modules. Their system is not based on an external clock (as would be the case in electronics) but on the existing machinery of the host cell, which acts as a synchronization device. In a highly informative article in the same issue of *Nature*, Solé and Macía (2014) emphasize the success of the strategy. They suggest that Prindle's simple approach, which does not require complex engineering given the authors' wise choice to tinker with the cell's own components, paves the way towards using the cell's natural machinery to integrate multiple synthetic components in a simple yet robust way. Although this pragmatic approach to engineering life is in sharp contrast with the orthogonality assumed for synthetic biological circuits, it may open up a new scenario. In this new setting, overlapping natural biological circuits are not seen as a hurdle, but as an opportunity to do more engineering work with less complexity to cope with.

5.4 Cooling Down the Cool Engineer

Paradoxically, one of the major breakthroughs in synthetic and systems biology got almost unnoticed. Luis Serrano and coworkers in the Center of Genomic Regulation (Barcelona) and Peer Bork and his team at the European Molecular Biology Laboratory (Heidelberg) published in 2009 three papers in the same issue of *Science* on the naturally reduced bacterium *Mycobacterium pneumoniae*. The authors used genome-scale screening for soluble protein complexes and estimated the number of molecular biomachines (some 200) and they finally concluded that *M. pneumoniae*, one of the simplest cellular organisms, exhibits a proteome complexity (an emergent one?) that, as they stated "could not be directly inferred from its genome

Fig. 5.1 Adventures in synthetic biology. A comic created by Drew Endy and Isadora Deese and the MIT synthetic biology working group, illustrated by Chuck Wadey and published in 2005 in *Nature* to explain the foundations of this discipline. Available for free in: http://mit.edu/endy/www/scraps/comic/AiSB.vol1.pdf (accessed 10 Apr 2014)

composition and organization or from extensive transcriptional analysis" (Kühner et al. 2009). In the second article, the group characterised reactions catalyzed by 129 enzymes by means of a notable, titanic work (more than 1,300 growth curves), revealing, the complexity in the form of metabolite concentrations, cellular energetics, adaptability, and global gene expression responses (Yus et al. 2009). One of the finest results of this work was the demonstration of the predictive power of the metabolic model for the cell. Despite their intracellular lifestyle, *Mycoplasma* cells can be grown under lab conditions using an extraordinary complex culture medium. Scientists were able to define the minimal set of metabolites for growing *Mycoplasma* cultures with the remarkable result of better growth kinetics using a synthetic, theoretically deduced minimal medium. Finally, in-depth transcriptome analysis of *M. pneumoniae* revealed, once again, an unexpected complexity (Güell et al. 2009). It is interesting to highlight that these reports, along with reports by other Old World researchers, are typical examples of what one could call the European approach to synthetic biology: a precautionary, high-throughput characterization-based strategy, which does not presuppose the engineering, machine-like nature of living cells.

We recognize, together with the historian Luis Campos, that maybe there is a European way of doing and viewing synthetic biology, influenced by specific cultural and institutional contexts. Instead of considering it a "foundational technology" synthetic biology is approached by many European scientists with a broader and flexible view, more as an "empowering of existing [scientific and technological] interfaces" (Víctor de Lorenzo's opinion in Campos 2013).

5.5 Biofuels

Another less-known hot topic on synthetic biology is that of the production of biofuels. Along with development of the discipline, several reports have announced high yields of microbial production of alcohols, fatty acid alkyl esters, alkanes, and terpenes. Ideally, third-generation biofuels are expected to be obtained from renewable resources such as non-edible lignocellulosic agricultural or forestry residues (e.g., wood, grass, straw), urban residues or oil wastes. Complex sugars from these have to be first converted into simple sugars, which are in turn converted into biofuels during the anaerobic growth of microorganisms. Metabolic intermediates of central (glycolysis, fatty acid biosynthesis) or peripheral pathways (e.g., 2-keto acid and isoprenoid pathways) are the starting materials for the synthesis of several biofuel components, like isobutanol, alkanes or sequiterpenes (Fig. 5.2). Advances in a sustainable and optimal industrial use require synthetic-biology approaches aimed not only to optimize biological hosts for pathway use (i.e., proper consumption of feedstock and maximum metabolic yields) but also for engineering the resistance to both inhibitors derived from the original biomass and their own fermentation products. Albeit some notable successes have been reported, there are still enormous obstacles for the substitution of petroleum-based fuels by advanced biofuels (for comprehensive reviews see Peralta-Yahya et al. 2012; Huffer et al. 2012; Buijs et al. 2013; Wen et al. 2013).

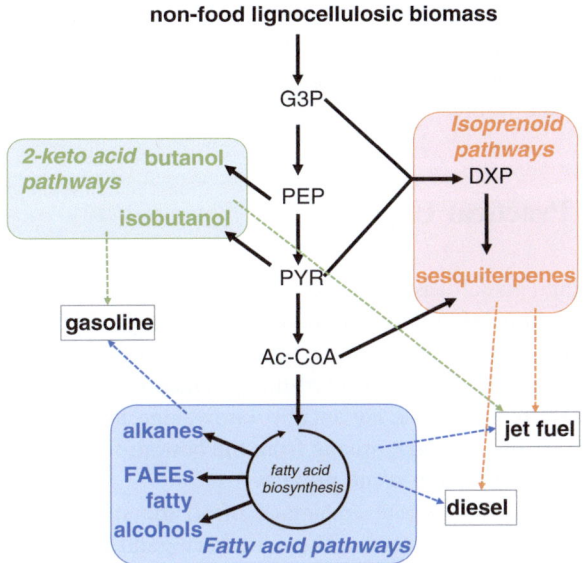

Fig. 5.2 Metabolic origin of third-generation biofuels. Ideally engineered microorganisms will be able to transform non-food lignocellulosic waste into useful biofuel components: isobutanol and other higher alcohols for gasoline, isobutanol and sesquiterpenes for jet fuel, and FAEEs, sesquiterpenes and alkanes for diesel. Abbreviations: G3P, glyceraldehyde-3-phosphate; PEP, phosphoenolpyruvate; PYR, pyruvate; Ac-CoA, acetyl Coenzyme A; DXP, deoxyxylulose-5-phosphate; FAEEs, fatty acid ethyl esters. Adapted from Huffer et al. (2012), Buijs et al. (2013)

5.6 Armpit Cheese or Public-Oriented Research in the Name of Synthetic Biology

Serrano and Bork's work is an example of an impressive research project that will strongly influence the synthetic-biology scientific community but it will have almost no impact on the public. By contrast, some heterodox synthetic biology projects at the interphase between science, arts and public communication are examples of the opposite. For instance, Stanford University and the University of Edinburgh collaborated with young researchers who prepared "human cheese" from human-inhabiting bacteria. This was one of the most popular activities within the project Synthetic Aesthetics,[1] in which a group of synthetic biologists, designers, artists and social scientists are working together "to explore collaborations between synthetic biology, art and design".

What they do is not standard research on synthetic biology, but an exercise of critical exploration of the discipline. Indeed, their creative search for novelty is shared by scientific goals. Additionally, although their provocative work was not

---

[1] http://www.syntheticaesthetics.org/ Accessed 10 Apr 2014. See also Ginsberg et al. (2014).

conceived to make a highly technical—and controversial—science likeable to the public, time will tell whether such unorthodox approach to our relationship with microorganisms will contribute to making society less reactionary to genetic modifications in the future.

5.7 Beyond Practical Uses

It would be an intellectual waste to ignore the scientific fruits provided by synthetic biology, in sole favour of potential short-term technological benefits (Ruiz-Mirazo and Moreno 2013). This is the opinion held by some of synthetic biology's most outstanding leaders, from the most hard-line engineering schools. Their confrontational statements belittling the advance in knowledge that synthetic biology can provide (especially, but not only, arising from the bottom-up approach) demonstrate the intellectual sloppiness that some profess.

It would seem smarter to take advantage of all the strategies available and to review a little history along the way. There are very good examples of what can be learned through synthesis, inaccessible to us through observation and analysis alone. One of the best examples may be the case of the chemical synthesis of vitamin B12, for which the Woodward-Hoffmann rules on the symmetry of molecular orbitals were established.

Everyone assumes synthetic biology will be hugely successful within biotechnological or biomedical ambits. But, as Benner (2003) and Szathmáry (2004) pointed out, hopefully it will indeed be a success if it sheds light on some of the more fundamental aspects of the essence of life, including new and universal ideas on biology that remain inaccessible using simply analytic tools.

References

Benner S (2003) Synthetic biology: act natural. Nature 421:118

Buijs NA, Siewers V, Nielsen J (2013) Advanced biofuel production by the yeast Saccharomyces cerevisae. Curr Op Chem Biol 17:480–488

Campos L (2013) Outsiders and in-laws: Drew Endy and the case of synthetic biology. In: Harman O, Dietrich MR (eds) Outsiders scientists: routes to innovation in biology, Chap. 18. University of Chicago Press, Chicago, pp 331–348

Chan LY, Kosuri S, Endy D (2005) Refactoring bacteriophage T7. Mol Syst Biol 1(2005):0018

Endy D (2005) Foundations for engineering biology. Nature 438:449–453

Gibson DG, Benders GA, Andrews-Pfannkoch C et al (2010) Creation of a bacterial cell controlled by a chemically synthesized genome. Science 329:52–56

Ginsberg AD et al (2014) Synthetic aesthetics. Investigating synthetic biology's designs of nature. MIT Press, Cambridge

Güell M, van Noort V, Yus E et al (2009) Transcriptome complexity in a genome-reduced bacterium. Science 326:1268–1271

Huffer S, Roche CM, Blanch HW, Clark DS (2012) Escherichia coli for biofuel production: bridging the gap from promise to practice. Trends Biotech 30:538–545

Kühner S, van Noort V, Betts MJ et al (2009) Proteome organization in a genome-reduced bacterium. Science 326:1235–1240

Lartigue C, Vashee S, Algire MA et al (2009) Creating bacterial strains from genomes that have been cloned and engineered in yeast. Science 325:1693–1696

Macía J, Posas F, Solé R (2012) Distributed computation: the new wave of synthetic biology devise. Trends Biotech 30:342–349

Peralta-Yahya PP, Zhang F, del Cardayre SB, Keasling JD (2012) Microbial engineering for the production of advanced biofuels. Nature 488:320–328

Prindle A, Selimkhanov J, Li H et al (2014) Rapid and tunable post-translational coupling of genetic circuits. Nature 508:387–391

Ro DK, Paradise EM, Ouellet M et al (2006) Production of the antimalarial drug precursor artemisinic acid in engineered yeast. Nature 440:940–943

Ruiz-Mirazo K, Moreno A (2013) Synthetic biology: challenging life in order to grasp, use, or extend it. Biol Theor 8:376–382

Solé R, Macía J (2014) Synthetic biology: biocircuits in synchrony. Nature 508:326–327

Szathmáry E (2004) From biological analysis to synthetic biology. Curr Biol 14:R145–R146

Wen M, Bond-Watts BB, Chang MC (2013) Production of advanced biofuels in engineered *E. coli*. Curr Opin Chem Biol 17:472–479

Young L, Noskov VN, Glass JI et al (2008) Complete chemical synthesis, assembly, and cloning of a Mycoplasma genitalium genome. Science 319:1215–1220

Yus E, Maier T, Michalodimitrakis K et al (2009) Impact of genome reduction on bacterial metabolism and its regulation. Science 326:1263–1268

Chapter 6
The iGEM Competition

Abstract The international Genetically Engineered Machine (iGEM) competition is a well-known example of synthetic biology and a workbench for the development of heterodox, multidisciplinary and frontier work made by undergraduate students. We review the origin, organization and structure of the competition; we describe how an iGEM team can be set in place, and briefly summarize some of the main milestones and challenges of a competition that is only one decade old. We discuss the links of the competition with the Registry of Standard Biological Parts and the flagship role of iGEM as a very trench of the synthetic biology revolution.

Had we to choose one example to define the very spirit of synthetic biology, it would be the international Genetically Engineered Machine (iGEM) competition. A search on Google trends shows the interest in the competition as deduced from internet searches, with a seasonal profile characterized by annual peaks around November, when the World championship takes place. The first thing we should point out about iGEM is the obvious fact that it is a competition. Thousands of students worldwide spend a long summer in an academic lab trying to set up a sound biotechnological project combined with surprisingly complex mathematical modelling, dissemination activities and social sciences reports. The project is rounded off with a poster, an almost professional wiki, and an oral presentation, the complexity of which is in sharp contrast with the youth of the synthetic biology apprentices responsible for each project (Fig. 6.1).

Both authors of this book have been involved in iGEM for 5 years as supervisors and we were stunned when we first learnt the details of this outstanding exercise of organization, high-level biotechnology and show business. iGEM is both a good example of ingenious scientific approaches and an exciting way to promote high level research among science students. In this chapter, we will describe the close links between the competition and the synthetic biology philosophy, give an overview of the historical evolution of iGEM, describe its structure and phases, and give some details of several selected iGEM projects. Finally, we will try to define the boundaries of the competition and the challenges it faces if it is to keep its current role as a media-oriented example of the potential of synthetic biology approaches.

© The Author(s) 2014   55
M. Porcar and J. Peretó, *Synthetic Biology*,
SpringerBriefs in Biochemistry and Molecular Biology,
DOI 10.1007/978-94-017-9382-7_6

Fig. 6.1 Geographical distribution of attendance, awards and medals in iGEM editions. This interactive map can be accessed at iGEM.org. The picture shows geographic distribution throughout the iGEM history (2004–2013). From iGEM.org accessed 10 Apr 2014

6.1 iGEM—Synthetic Biology for the Youngest Scientists

A primitive version of iGEM was born in 2003 as an Independent Activities Period (IAP) of the Synthetic Biology Group at MIT, founded by Tom Knight, Drew Endy and Randy Rettberg. In 2004 iGEM was launched with the participation of just five teams (the "i" of iGEM meant "intercollegiate" and not "international" at that time). As of 2006, iGEM became truly international and started to grow; a trend that continues today, with around 200 teams attending the four regional jamborees.

It is not fortuitous that one of the most important undergrad competitions revolves around synthetic biology. In fact, to a certain extent, iGEM could be seen as a thermometer of developments in SB, as this thriving competition correlates with synthetic biology prospering as a discipline. Likewise, we believe that if the competition were to decline in the future, so would the very discipline. The rationale behind this statement is simple: the notion of life engineering in synthetic biology aims to revolutionize biotechnology by applying engineering principles. Thus, synthetic biology-based biotechnology is expected to be relatively easy compared to traditional trial-and-error biotechnology. So easy that, with moderate supervision and material support, even undergrad students should be able to carry out ambitious projects and successfully construct genetically modified cells, in just one summer. The very existence of iGEM is a consequence of this spirit, in which we provide bright young people with standards and tools and let them get on with engineering life (Fig. 6.2).

Fig. 6.2   iGEM from above. A picture of iGEM 2013 finalist teams at MIT. http://2013.igem.org/ Welcome. Accessed 10 Apr 2014

6.2   From Biobrick to Jamboree

Above, we have outlined the origins of iGEM and its central role as an experiment on the viability of synthetic biology. Now, we will describe how an iGEM edition works. The best way to do so is to start by explaining the paradigm of the iGEM working unit: the Biobrick^{TM} part. Indeed, to be eligible to win a medal in the competition, competitors must submit a minimum of one Biobrick part to the Registry of Standard Biological Parts, although most teams submit dozens and some register hundreds. BioBrick parts are defined as DNA sequences of defined structure and function; designed to be incorporated into living organisms—such as bacterial cells—in order to construct new biological systems (Carlson 2011). They can be either natural (mimicking a naturally occurring DNA sequence) or engineered (gene fusions or important modifications of the protein-encoding sequences). Biobrick parts are assumed to be standard and, thus, module-ready. They can be used alone or combined with other parts to yield a higher level of complexity. These more complex levels include biological devices, which can—in turn—be combined to form biological circuits working—in principle—orthogonally inside the host cell. Their standard and modular nature is claimed to be a consequence of the prefix and suffix adapters they bear, which are simply short polylinker-like sequences with several restriction enzyme sites, similar to plasmid multicloning sites. If we use enzymatic digestion and ligation appropriately, we can rapidly clone a desired Biobrick part, often designed from a natural source by a Polymerase Chain Reaction (PCR) with prefix- and suffix-containing specific primers. It should be noted, though, that the ease with which Biobrick parts can be combined to form more complex structures does not necessarily imply that they behave in a standard and orthogonal way; the design of Biobrick parts assures a modular construction, but the desired behaviour has to be experimentally determined on an ad hoc basis. In other words, there is nothing in the way Biobrick parts are assembled that assures their functional compatibility with other Biobrick parts, devices or the host cell metabolism, all of which, in last instance, rely on the information contained in DNA sequences and not necessarily in the way they are physically assembled. At the end of this chapter and in the last chapter of the book we will discuss both the potential and the limitations of the use of standards in iGEM in more detail and, by extension, in synthetic biology.

But now let's go back to the challenge of setting up an iGEM project. How is an iGEM team organized? How do students find supervision or financial support? Is it expensive to participate? Could you, the reader, interested in artificial life, become the next *igemite* to win the most coveted golden Biobrick trophy?

In practice, iGEM teams are born, organized and supervised by academic institutions (typically one university or a *coalition* of several local universities). One staff member—the instructor—is the official contact person and he/she is usually the general coordinator of the project. Students are often attracted by news reports, lecturers informing them about iGEM or by veteran participants. Many institutions run dissemination activities aiming to attract students' attention. Preferred backgrounds are biotechnology and engineering, but a plethora of other studies and skills, ranging from artistic or social to web design are commonly found among iGEM students. When too many students apply for a "position" in an iGEM team, a selection process might be set in place to ensure a balanced team (typically 6–12) of highly motivated students. In our experience, besides the ability to work in a team, the most valuable skills are creativity, endurance, enthusiasm and English proficiency. In our "castings", we try to detect students with these abilities and specific knowledge in biotechnology, modelling, human practices or informatics, in order to combine these "human modules" in a robust "host" team. Interestingly, modularity, decoupling and black-box design, which are typical features of synthetic biology-based biotechnology, can already be found the iGEM teams' composition.

Teams are often subdivided in subteams. For example, in 2013 at Valencia Biocampus (Fig. 6.3), we supervised a research project, *Wormboys*, aiming to establish an artificial symbiosis between the nematode *Caenorhabditis elegans* and two bacterial species: *Pseudomonas putida* and *E. coli*.[1] For practical reasons, we organized our 10 students in sub-teams of 2–3 people each: wiki, wetlab, drylab (hardware and modelling), poster, presentation, Biobrick parts and Human Practices (an odd term used almost uniquely in the iGEM context that refers to ELSIs, Ethical Legal and Social Implications of science, in this case, synthetic biology). Each student typically participates in 2–3 different sub-teams, which gives all members of the team a wider overall vision of the project. We do not know how other teams choose their project topic, but the close relationship between the field of research

---

[1] For further details visit the wiki of Valencia-Biocampus 2013 team, *Wormboys* project: http://2013.igem.org/Team:Valencia_Biocampus.

pursued by the group leaders and their iGEM project would suggest the election of a suitable—i.e., feasible—topic is "guided". In our case we choose the topic in a kick-off meeting with the participation of the whole team (roughly 20–30 people). Students are encouraged to present at least one project proposal in a short (5–10 min) talk and a general discussion follows in which, ideally, either one proposal or a combination of more than one are selected as the germ of the future iGEM project.

In principle, the structure of an iGEM team is simple. The students should be organized under the supervision of at least one general instructor. In practice, though, each team is hierarchically organized on two or three levels, for instance, the core of students work in the lab with senior researchers supervising their work. However, in many cases, there are three levels: senior instructors, students, and an intermediate level of advisors (typically, PhD students, post-docs and/or veteran igemites). Once the project is in place and sub-teams have been created, students start working on their respective topics under the supervision of expert instructors and advisors in each field. One of the most enriching aspects of iGEM is that students share the same project, whereas supervisors of wiki and Human Practices do not have the same global view but rather perform a specialised role.

Now we should imagine a research institute with most of its staff on vacation but one or two laboratories buzzing with very young researchers doing wetlab experiments, modelling or preparing a poster for the competition. They may well be stressed, working towards deadlines. After a few months of work, wikis freeze a couple of weeks before the competition. This means that all the work, with the exception of the poster and the oral presentation, has to be done and correctly reported on the website. After wiki freeze, which happens by mid September, students tend to concentrate on their studies (though some of them may be busy with the presentation) until the first phase of the competition arrives: the regional jamboree. After 2010 as the sheer number of students attending the iGEM made a single jamboree unmanageable, the competition was split into two phases: a regional one (in 2103: North America, Latin America, Europe and Asia) and a final World Championship at the MIT headquarters in Cambridge, Massachusetts. The regional phase acts as filter to the World phase, to which around just one third of the teams will pass.

Both the regional and the World championships usually take the form of 3 days of meetings. The first is devoted to registration, the second is the core of the competition with all the presentations and most judging activities, and on the last day the awards ceremony takes place, with the presentation of the finalists (usually three). In 2014 an extraordinary Giant Jamboree will take place in Boston, with no regional phases—just the final phase. It is not clear yet whether this will be in 2014 alone or whether it will be the new iGEM competition format.

Awards at iGEM are divided into medals and special awards. All teams are eligible for a medal (bronze, silver or gold) provided that they meet a list of criteria. This system means all teams have a fair chance of obtaining an award according to

Fig. 6.3 Valencia Biocampus iGEM team in summer 2013. From *left* to *right*, *standing*: Alba Corman, Alba Iglesias, Alejandro Valero, Pedro Dorado, Alejandro Torres, Jessica de Loma; in the *forefront*, Marina Mañas, Guillermo Zafrilla, Tonny Ruiz and Samuel Miravete. Picture by Miguel Lorenzo/University of València

their accomplishments, without competing directly with each other. The special awards (i.e., best wiki, best poster or best experimental measurement) represent the real competitive element, there being just one per Jamboree. Regarding promotion, only a small number of teams from each region can advance (1/3 in 2012). Selection is made by picking the top-scoring teams, and the rate of selection is proportional to the number of teams from a particular region relative to the total number of teams in that particular iGEM edition.

6.3 The Outcome and the Future

Of the thousands of iGEM projects presented so far, only a few have been published but some of them are noticeable examples of engineering ingenuity (Vilanova and Porcar 2014). For instance, in 2004 the UT Austin team made a real breakthrough with the first bacterial photograph, which they developed using a biofilm that could detect edges in projected images through the control of gene expression in *E. coli* using light. It was published one year later in *Nature*. Another outstanding iGEM project involved the design of a DNA-linked metabolic network, which beautifully reproduced chain production by physically placing enzymes on the DNA strand one after the other according to the order in the pathway, presented by the Slovenia team in 2008 (i.e., scaffolding, see Chap. 4). Then, in 2009, we were delighted to see our team awarded the Second Runner-up, Best New Application and Best Experimental Measurement with an unorthodox project aiming at making what we called iLCDs, iGEM Living Cell Displays, with electrically stimulated aequorin-expressing yeast

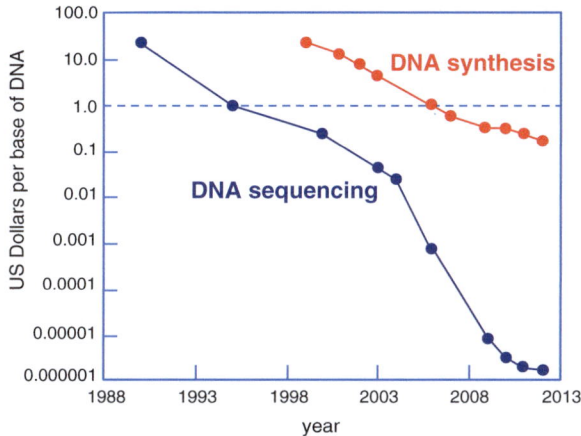

Fig. 6.4 Cost of DNA sequencing and synthesis. Note the ten-year gap between the onset of both technologies and the similarities in both *curves*. If recently announced new technologies lower the cost of chemical synthesis, it is expected that, by 2020, the cost of DNA synthesis will be five times lower than it is today. (Adapted from R. Carlson <www.synthesis.cc> accessed 10 April, 2014)

as living pixels. The same year, Cambridge was awarded the Biobrick trophy for a project entitled "*E. chromi*", a new development in which a spectrum of pigments were expressed in *E. coli*, designing a set of Colour Generators. Many more topics have been attempted by iGEM teams: bioremediation with engineered strains, sensing devices, (bio)logical switches, cells engineered to count, bacteria-based food quality assessment… the list of astounding iGEM projects is almost endless.

At this point, it is important to point out that things in iGEM may not be as easy as they look. From our experience as supervisors of the Valencia and Valencia Biocampus teams, we know that Biobrick-based ligation is sometimes as tedious as standard molecular cloning—if not more so—, and synthetic constructs tend to either not work at all, or to be very unstable. Orthogonality or independent behaviour should not produce side effects when transforming host cells with vectors containing synthetic constructs. But orthogonality is more an exception than a rule. The performance of the thousands of Biobrick parts submitted to the registry is uncertain and, the fact is, most teams tend to design and use their own Biobricks rather than choosing those from the Registry (Vilanova and Porcar 2014). This lack of trust in Biobrick parts from the Registry is in stark contrast with the philosophy underlying the competition. To make things worse, the dropping costs of DNA chemical synthesis could herald the death of the Biobrick-based assembling system (Fig. 6.4). If it is easier and cheaper to order whole constructs, why bother to build individual parts and then combine them? Certainly, future iGEM editions will have to co-evolve with both the available technological advances and the current opinions and sensitivities in such diverse field as synthetic biology.

References

Carlson RH (2011) Biology is technology. The promise, peril, and new business of engineering life. Harvard University Press, Cambridge

Vilanova C, Porcar M (2014) iGEM 2.0: refoundations for engineering biology. Nat Biotech 32:420–424

Chapter 7
Are We Doing Synthetic Biology?

Abstract Even for synthetic biologists, it is not obvious what is and what is not synthetic biology. The interphase between biotechnology and metabolic engineering is still large and complex. Although synthetic biology is expected to rely on modelling and rational design, the fact is that synthetic constructs are still prone not to work as expected. This makes arguable the "genome writing" era some already claim we are at. Due to the incomplete knowledge of the complexity of the living we are not at the writing but only slightly ahead of the copying era. An analysis of the use of the term "synthetic biology" as well as the trends exhibited by the iGEM competition strongly suggests a more cautious use of synthetic metaphors.

In this book, we have reviewed synthetic biology, an emerging research field with (at least) a dual definition encompassing both the design and construction of new biological parts, devices and systems, as well as the re-design of existing parts for useful purposes. As we have stressed, this framework is vague enough to include modern biotechnology and many other connected fields (Fig. 7.1). However, synthetic biology is generally considered to differ from biotechnology, metabolic engineering or systems biology (Porcar and Peretó 2012). It is the engineering, systematic and design-based facet of synthetic biology that sets it apart. Its ultimate goal is to make life easier to engineer or, at least, "an attempt to make biology less qualitative and descriptive and more quantitative and predictive".

Synthetic biology's great expectations include: mass production of an array of useful compounds, from drugs to biofuels; a key role in the development of bioremediation; dramatic increases in crop yield with the production of novel food ingredients and a large variety of chemicals; and improvements in medicine.

A search run in the *Scopus* database for scientific documents whose titles, abstracts or keywords include "synthetic biology" reveals there has been an exponential growth the use of these terms in the last 10 years (Fig. 7.2).

However, the question remains open as to whether this exponential growth of synthetic biology in the literature corresponds to an actual revolutionary expansion of this discipline or just to a fashionable use of the term, undoubtedly related to the

M. Porcar and J. Peretó, *Synthetic Biology*,
SpringerBriefs in Biochemistry and Molecular Biology,
DOI 10.1007/978-94-017-9382-7_7

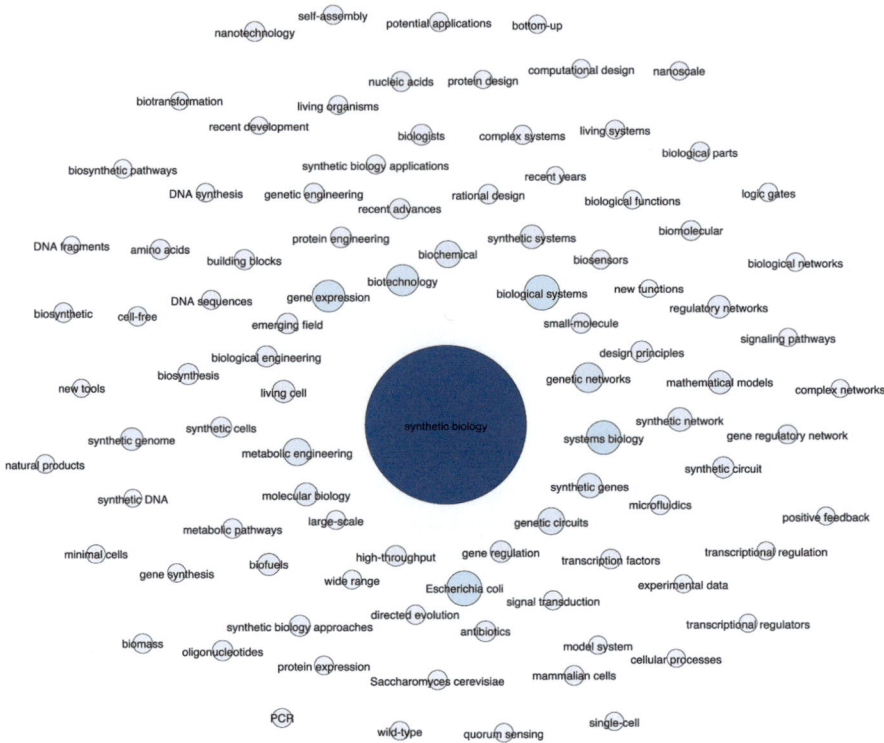

Fig. 7.1 The synthetic biology solar system? The figure shows a Fruchterman-Reingold representation in Gephi of the top synthetic biology-aggregated terms on titles, abstracts and keywords (Oldham et al. 2012)

fuzzy boundaries between synthetic biology and related disciplines (Fox Keller 2009). So, what is and what is not synthetic biology? One rule of thumb is the degree of sophistication. While we are still waiting for synthetic biology to achieve the high expectations it has set for itself, the milestones of discipline are assumed to exhibit a characteristically higher degree of "sophistication" compared to sister disciplines like metabolic engineering. Is this assumption supported? As we have described in Chap. 5, in 2006 there was a report on how the precursor of artemisinin had been produced in engineered yeast, and then in *Escherichia coli* (Keasling 2010). This achievement rested on engineering the native mevalonate pathway, inserting two genes from *Artemisia annua,* the plant from which "natural" arte-misinin is extracted. This work, which was further improved and helped bring down the costs of malaria therapies worldwide, is generally considered one of the flag-ships of synthetic biology. However, it is difficult to draw a line defining the bounds of sophistication or to explain why some notable metabolic engineering examples, such as that of carotene (pro-vitamin A) synthesis in the so-called Golden Rice, are not commonly considered as canonical exercises in synthetic biology.

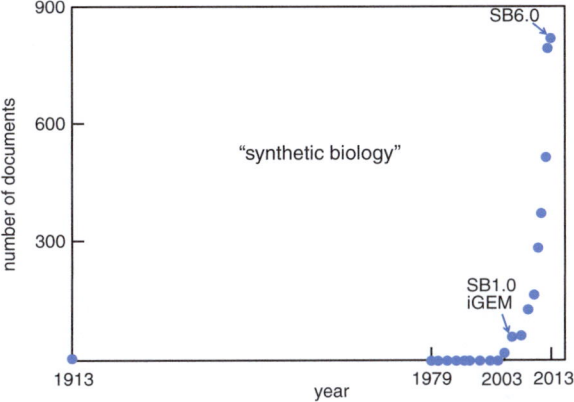

Fig. 7.2 Results of a Scopus search for the terms "synthetic biology" in the title, abstract or key words of documents during the last 100 years. The 1913 reference is a book review on Leduc's Synthetic Biology in the journal *Nature*. From 1979 to 2002 we retrieved 12 documents essentially related to genetic engineering or artificial life. In January 2003, in a comment in *Nature* Steven Benner (Benner 2003) used the term "synthetic biology" with one of its current values. Few months later, Tom Knight released the first MIT Synthetic Biology Working Group Technical Report on the idea of standardized biological parts (Knight 2003). The first iGEM contest and the first international meeting on synthetic biology (SB1.0) were held in 2004 at MIT

7.1 A Word on Genomes: Are We True Writers?

As we have seen in this book (Chap. 5), one of the most sound technical break-throughs was the synthesis of the so-called first synthetic cell (or even "Synthia"); a synthetic chromosome-driven cell with natural components. The term genome writing, or even life writing has been used since the report published by J. Craig Venter's team in *Science* in 2009. Once again in synthetic biology, the metaphor might be an overstatement (Giuliani et al. 2011; Bedau et al. 2010). It is true that genomic information can be compared to a book. But, we should ask ourselves whether, strictly speaking, copying information without significant changes is true writing. We refer the reader to Fig. 7.3, which shows a more accurate metaphor of a synthetic "text" compared to an original wild-type one.

At the time we were completing this book, another report on "synthetic chro-mosomes" appeared, this time referring to an artificially synthesized yeast chro-mosome: synIII is a shorter version of yeast chromosome III (Annaluru et al. 2014). The sequence was significantly simplified (by removing mobile elements, introns or some redundant, non-essential genes) and modified (addition of artificial telomeres, changes in some codons, and insertion of tags to differentiate the artificial version from the natural one), but the global genetic and—of course—metabolic architec-tures of the cell were kept without any apparent loss of fitness. In addition, the synthetic chromosome is equipped with molecular devices that will make it possible

Good now, sit down, and tell me, he that knows,
Why this same strict and most observant watch
So nightly toils the subject of the land,
And why such daily cast of brazen cannon,
And foreign mart for implements of war;
Why such impress of shipwrights, whose sore task
Does not divide the Sunday from the week;
What might be toward, that this sweaty haste
Doth make the night joint-labourer with the day:
Who is't that can inform me?

Good now, sit down, and tell me, he that knows,
Why this same strict and most observant watch
So nightly toils the subject of the land,
And why such daily cast of brazen cannon,
And foreign mart for implements of war;
Why such impress of shipwrights, whose sore task
Does not divide the Porcar and Peretó Sunday from the week;
What might be toward, that this sweaty haste
Doth make the night joint-labourer with the day:
Who is't that can inform me?

Fig. 7.3 Writing versus plagiarizing genomes. The complete synthetic sequence of an artificial—yet functional—bacterial chromosome has been proposed as the first step towards "writing" genomes (see Chap. 5). However, analogous to Richard Dawkins' "Blind Watchmaker" we are still DNA "blind writers" or "printers" rather than true authors: *on the left*, Marcellus' speech from *The Tragedy of Hamlet, Prince of Denmark* (Act 1, Scene 1), by William Shakespeare; on the right, pasted copy of the same text with a few artificially inserted, out-of-frame characters

to further reduce, reorganize or modify at will the sequence, potentially implementing new functional properties.

Now a fully synthetic eukaryotic chromosome is technically feasible. Shall we then be "eukaryotic genome writers"? The fact is, though, that genomes are composed of genetic information in the form of DNA sequences, not unlike real writing, but the main difference between these two codes is that the function of most DNA sequences forming a genome are either unknown or poorly understood, or they code for proteins whose functions are completely unknown or not well characterized: even for a well studied microorganism like *E. coli* one-third of the proteome remains functionally uncharacterized. True writers work with very well characterised parts—words–, which interact with each other in fully predictable ways—grammar—to yield increasingly complex building blocks—sentences, paragraphs, this book. Without in-depth knowledge of the function of biological blocks and how they interact, it is a hazardous over-simplification to equate writing concepts such as *words, grammar, sentences, paragraphs* and *books* with, respectively, *genes, metabolism, devices, circuits* and *synthetic organisms*.

Thus, we believe that use of the term "writing" should be restricted to intentional synthesis of fully characterized artificial DNA assemblies demonstrating predictable behaviour, and not to chemically synthesized, essentially verbatim copies of existing genomes.

7.2   Is Life Engineerable?

One of the most influential reports on synthetic biology represents both a milestone of the rational design of living systems and an example of the limitations of this approach. As we have seen in *refactoring phage T7* (Chap. 5), a pioneer approach aiming to make a virus behave "more logically", it is in fact concluded that the vast majority of the rationally introduced genomic rearrangements yielded lower biological fitness (since they made smaller lysis plaques) than the wild-type strains.

The other examples analysed in this book, microbial synthesis of artemisinin or the chemical synthesis of a functional chromosome, all reveal how hard it is to rationally design life, and the continued need for tinkering and fine tuning of biological systems. As stressed by Herbert Sauro in a special issue on synthetic biology appeared when we were completing the edition of this book: "Bioengineers painstakingly craft a design, and a day later it has crumbled in the face of evolutionary selection" (Collins et al. 2014). One of the reasons behind this need is the key issue of biological complexity and messiness. The famous quotation by Drew Endy on emergent properties (Chap. 1) reveals the unease engineers feel when they approach the complexity of living things. To reduce this fuzzy complexity, engineering principles of standardization, decoupling and abstraction should be so powerful that even undergrad students can develop relatively complex synthetic biology projects. And this is exactly what iGEM is about. But, in what extension are the engineering pillars applicable to life?

7.3   Standards in Biology: The iGEM Competition

The iGEM competition (see Chap. 6) has a very simple goal: to educate the next generation of synthetic biologists by allowing them to build simple biological systems from standard, interchangeable parts. This is achieved by providing each team with a library of standardized (Biobrick^{TM}) parts that on paper allow infinite combination of biological parts in a cell, in a Lego-like fashion. Interestingly, iGEM teams can either use Biobrick parts from a repository (The Registry of Standard Biological parts) or, alternatively, they can submit new Biobrick parts to the Registry. Successful, award-winning projects often avoid using previously characterized standard parts from the Registry and choose to design, characterize, use and submit new "standards" to the registry. This apparent paradox reveals the competitors' doubts about using in a particular biological system parts developed by other teams (Vilanova and Porcar 2014).

The authors of this book have conflicting feelings about iGEM. As an educational experience, the competition is simply priceless. Moreover, it is difficult to imagine a better way to inculcate students with values like excellence, hard work, integrity and enthusiasm: iGEM values. It has to be stressed, though, that enthusiasm is not a goal but the way towards achieving scientific excellence. As iGEM enthusiasts ourselves, and having attended the competition as instructors or advisors for years, we want to emphasize the worrying trends in this competition, which should evolve towards a different approach regarding the concept of standard. Firstly, a clear distinction between building (prone to physical assembly) and functional (universal behaviour) standards should be made. And secondly, a critical review of malfunctioning biological standards (both inside and outside the iGEM literature) should be tackled. The very concept of biological standard is not to be taken for granted. In engineering, standards are conceived with boundaries such as tolerances (unintended deviations from the designed shape/size) and allowances (intentional deviation to allow, for example, an imperfect shaft to fit an also imperfect hole). In synthetic biology, though, this concept has only recently been proposed. It seems reasonable that combining rational design with flexibility in the form of directed evolution, or including biological allowances could help to booster the discipline.

The coming years will be exciting, we will see entirely synthetic bacterial and eukaryotic genomes, incredibly cheap DNA synthesis; a dramatic increase in the knowledge of biological complexity from a systems perspective–from metabolic networks to bacterial consortia–, and the development of a far more reliable toolbox to genetically engineer life. Learning more on life through systems and synthetic biology approaches will smooth the way to a predictable engineering of life. Between fear and hype, synthetic biology, now in its infancy, might be the main technological force with which to face the huge environmental and medical challenges of the 21st century.

References

Annaluru N, Muller H, Mitchell LA et al (2014) Total synthesis of a functional designer eukaryotic chromosome. Science 344:55–58

Bedau M, Church G, Rasmussen S et al (2010) Life after the synthetic cell. Nature 465:422–424

Benner S (2003) Synthetic biology: act natural. Nature 421:118

Collins JJ et al (2014) Synthetic biology: how best to build a cell. Nature 509:155–157

Fox Keller E (2009) What does synthetic biology have to do with biology? BioSocieties 4:291–302

Giuliani A, Licata I, Modonesi CM, Crosignani P (2011) What is artificial about life? Sci World J 11:670–673

Keasling JD (2010) Manufacturing molecules through metabolic engineering. Science 330:1355–1358

Knight TF (2003) Idempotent vector design for standard assembly of biobricks. MIT Synthetic Biology Working Group Technical Reports. dspace.mit.edu/handle/1721.1/21168. Accessed 10 Apr 2014

Oldham P, Hall S, Burton G (2012) Synthetic biology: mapping the scientific landscape. PLoS ONE 7:e34368

Porcar M, Peretó J (2012) Are we doing synthetic biology? Syst Synth Biol 6:79–83

Vilanova C, Porcar M (2014) iGEM 2.0: refoundations for engineering biology. Nat Biotech 32:420–424

Postface

Fictional stories, particularly those that try to forecast what the future will deliver to us in terms of science and technology, do not pass the test of time. Whether books of movies, we often look at those old creations with some degree of condescendence, fully aware of their failures or misjudgments. A Victorian novel, created by a young woman in the midst of a time of enlightenment, when science was starting to uncover the laws of nature and our own location within those laws, gives a remarkable exception to this rule. Two hundred years separates us from this book, which was written in the age of reason and translates some of the edge science of the time into a dramatic story. In her famous 1818 novel *Frankenstein*, an 18 years old Mary Shelley was able to create a character that has survived every single scientific advance, perhaps because the protagonist—a bright, but tormented Victor Frankenstein— encapsulates all the wonder and dangers of scientific breakthroughs. The result of Victor experiments is partially a failure: as soon as the creature gets awake, it turns out that it does not behave as expected. Eventually though, it becomes self-aware and looks (and gets) revenge.

It might appear as obvious that *Frankenstein* is mainly a story with a moral lesson: we shall not interfere with life nor modify it. There are red lines that cannot be crossed unless we cope with the consequences. But it is also a good picture of the far-reaching goals and intellectual challenges that predate Victorian science. It was known that electricity could make dead matter to behave as if alive. There was thus a technology that seemed capable of doing the unthinkable. Nowadays, Victor is remembered more as the scientist crossing the red line than the visionary. With the rise of genetic engineering and biotechnology, expectations raised among scientists whereas fears increased among non-scientists. We still experience today some non-rational consequences of those fears, which plagued many decisions involving genetically modified organisms. Now we have moved into a new level. The book by Porcar and Peretó provides us with a great synthesis of how the original ideas associated to the creation of life-like entities started and how it all ended up in a new engineered-oriented field: synthetic biology. The new field contains all the ingredients for leading us into a major change in our relationship with living systems.

M. Porcar and J. Peretó, *Synthetic Biology*,
SpringerBriefs in Biochemistry and Molecular Biology,
DOI 10.1007/978-94-017-9382-7

By looking at cells, tissues and organs as complex machines composed of multiple parts that we can modify, exchange or reconnect, bioengineers intend to modify the logic of these systems or subparts in order to get into novel functions that evolution has not generated, either because a different path was followed or because it is simply impossible to be obtained by natural processes. The first steps of the area reminds us those followed by other major technologies, such as microelectronics. It all started with the first large, simple and faulty transistors, which required considerable improvement before the age of combinatorial circuits fully developed, eventually leading to the revolution of information technology. Should we expect a similar outcome for synthetic biology? As discussed by the authors of this book, living matter differs from electronic systems in several ways, including among others the widespread cross talk that we observe between molecular parts used in designing circuits. This is no small difference, since we cannot extrapolate from microelectronic design rules into their living counterparts. But standardization of parts and a better understanding of its limits is developing and more complex circuits getting in place. Moreover, living systems include within them self-organization rules: many properties exhibited by biological structures result from emergence. Emergent properties are responsible for a big deal of the complexity displayed by living systems. They are obvious when we look at ant colonies, where individuals have no clue about how to build a nest or have a cognitive map of the environment. But the colony does. Similarly, little can be stored in terms of complex memories within a single neuron, but networks of connected neurons can. The presence of memories and their robust behavior cannot be explained in terms of single cells: we need the system and the interactions to get there. As discussed by the authors, emergent phenomena seem an undesirable part of the potential associated to synthetic biology. But this is actually not necessarily the case. Groundbreaking work involving stem cells and engineering efforts combining extracellular matrices and 3D printing revealed the potential of building tissues and organs almost for free. This shows that we might have a great deal of help coming directly from the ontogenetic rules that are already in place as a result of evolution. Building an eye as the engineers in Blade Runner might be a dream. Allowing stem cells to do their job and adding the appropriate engineering (some sort of self-organized engineering) can be in the future a major thread of this emerging field.

The field, as noted by Porcar and Peretó, is still in its infancy, but heading into adulthood at a good pace. As it has occurred with all previous technologies that eventually dominated our economics and society, it is moving through a phase of promise and hype. But soon will become part of the standard bioengineering that was a dream just a decade ago. In the process, we will surely learn a lot about fundamental biology. We will understand what can be modified and what cannot and, in doing so, uncover fundamental traits of the logic of living. This is already taking place under enormously creative teams of researchers coming from very different areas: biology, physics, chemistry, mathematics, engineering and computer science. It is actually interesting to remind that, in Shelley's story, Victor Frankenstein suggests at some point that any "man of science" should learn from all different fields. This is taking place now and perhaps not surprisingly an important part is done by young researchers not much older than Shelley herself. This time, however, reality might eventually win over the difficulties and limitations that pervade changing biological systems. Only time will tell us is we should worry about the consequences. Right now, there is room for hoping that we might successfully fight against diseases that have been resisting all kinds of treatments, improve and develop cheap vaccines and drugs that are required in places where they are needed, find ways of overcoming the challenge of climate change or help treating and preventing organ decay by novel ways of regenerating tissues. Surely this would have pleased Mary Shelley very much.

Ricard Solé
ICREA research professor
Complex Systems Lab (Institute for Evolutionary Biology, CSIC-UPF)
Barcelona

Author Index

© The Author(s) 2014
M. Porcar and J. Peretó, *Synthetic Biology*,
SpringerBriefs in Biochemistry and Molecular Biology,
DOI 10.1007/978-94-017-9382-7

Subject Index

© The Author(s) 2014
M. Porcar and J. Peretó, *Synthetic Biology*,
SpringerBriefs in Biochemistry and Molecular Biology,
DOI 10.1007/978-94-017-9382-7

MIX
Papier aus verantwortungsvollen Quellen
Paper from responsible sources
FSC® C105338

If you have any concerns about our products,
you can contact us on
ProductSafety@springernature.com

In case Publisher is established outside the EU,
the EU authorized representative is:
Springer Nature Customer Service Center GmbH
Europaplatz 3, 69115 Heidelberg, Germany

Printed by Libri Plureos GmbH
in Hamburg, Germany